Second Edition

Mechanical Fastening, Joining, and Assembly

James A. Speck

CRC Press
Taylor & Francis Group
Boca Raton London New York

CRC Press is an imprint of the
Taylor & Francis Group, an **informa** business

CRC Press
Taylor & Francis Group
6000 Broken Sound Parkway NW, Suite 300
Boca Raton, FL 33487-2742

First issued in paperback 2017

© 2015 by Taylor & Francis Group, LLC
CRC Press is an imprint of Taylor & Francis Group, an Informa business

No claim to original U.S. Government works

ISBN-13: 978-0-8247-5378-8 (hbk)
ISBN-13: 978-1-138-74840-8 (pbk)

Library of Congress Cataloging-in-Publication Data

Speck, James A., 1948-
    Mechanical fastening, joining, and assembly / author, James A. Speck. -- Second edition.
        pages cm
    Includes bibliographical references and index.
    ISBN 978-0-8247-5378-8 (alk. paper)
    1. Fasteners. 2. Joints (Engineering) 3. Assembling machines. I. Title.

TJ1320.S65 2014
621.8'8--dc23   2014035505

Visit the Taylor & Francis Web site at
http://www.taylorandfrancis.com

and the CRC Press Web site at
http://www.crcpress.com

Contents

Preface to the Second Edition

The second edition of *Mechanical Fastening, Joining, and Assembly* is presented as a guide and reference to all who engage in the mechanical arts. A decade and a half have passed since the preface to the first edition was written. This passage of time has affirmed many of the ideas expressed in that first edition. Time has also witnessed the development of some interesting, and for our purposes, intellectually important, technical progress in the assembly of products and in the design, manufacturing, and installation of fastener products and procedures. Time has not been idle even during periods when worldwide economic engines were moving at less than full throttle.

Ranging from the interest children exhibit in building things with toys such as interlocking blocks, rods, and connecting components to the advanced and sometimes strident adult voices and viewpoints expressed in the media, in academic institutions, and in political and public discourse around the world, a panoramic view of the subject of fastening and its foundational technologies, from the basics to the advanced is timely.

Research of technical education itself is undergoing at present a profound, and critically necessary, evaluation. And in the manufacture of products ranging from orthopedic hardware to aerospace, automotive, transportation, and defense products, the need to assemble and fasten components efficiently, and well into functional products of value, continues. In fact, as I write, the early reemergence of a growing U.S. manufacturing base can be witnessed in shops and manufacturing plants in large cities and small towns. We are making things! Knowledge and ideas build best when started from a solid foundation. What appears to me to be coming into focus is that manufacturing is a significant component of a nation's economy. From a strategic viewpoint, reliance on other industrialized economies for manufactured goods leads to dependence and potential weak points in an economy's self-reliance, even in a globally connected economy with many of today's economies. Manufacturing prowess provides strength and options not otherwise always available. A common issue I hear frequently in my business travels is the constraint of trained, young manufacturing employees. As those of us who are older reduce our hours of participation in manufacturing enterprises and begin to age and retire, a new workforce will be required. There is no doubt that this younger workforce has world-class computer and telecommunication skills. It is hoped that this will be matched by innovation in the expansion and growth of engineering skills to transform tomorrow's ideas and raw materials into value-producing, lifestyle-enhancing fastened products. I am optimistic.

In Chapter 1 of this edition, I have expanded the fastened components presented and reviewed. These are based on my experience in the field. At the end of each chapter is a review of engineering fundamentals with a focus on their application in the fastener industry. The first chapter concludes with a section on fastener statics. Using free body diagrams and learning to read the forces in a fastener and assembly are key skills. In Chapter 2, we take a deeper look at the fastener manufacturing processes, especially those I was exposed to in my career, cold heading and roll threading. Fastener dynamics is added in Chapter 2 to focus our attention on forces in motion, such as a wind turbine hub turning in strong winds. In Chapter 3, we take a look at materials in fastening and add fastener strength of materials. Somewhere in my fastener application engineering career I started to think of statics, dynamics, and strength of materials as the big three. Each of these serve as a key to good fastener applications work. Chapter 4 expands our review of the economics of fastening and provides some tools for engineering economics. Too often, technical people tend to steer clear of economic and financial factors. This does not have to be so. Neglecting economic and financial factors can lead to an over-reliance on technology to the detriment of the application and more importantly, the career advancement of the practitioner. Money is the life blood of an enterprise. In Chapter 4, we speak its language.

In Chapter 5, we examine the difference in static and dynamic strengths. These can sometimes present fastener application issues if not addressed properly. We also present some ideas with respect to chemistry in fastener work. Chapter 6 brings us a renewed look at modern assembly where the fundamentals still hold but with new constraints and concerns. Under the umbrella of numerical methods, we take an overview of the mathematical skills that provide the foundation of application engineering. Chapter 7 takes a look at fastener materials in this new century, provides some observations about the fastener laboratory, and discusses electrical theory. In Chapter 8, sustainability is focused along with some thoughts for application product management. A discussion of thermodynamics concludes Chapter 8 with material on energy systems and some new thought maps for application analysis. Chapter 9 provides the reader with a look at work from the fastener workshops that have been well received over the years and a look at a favorite application, D&D 100.

This second edition adds two new appendices. Appendix D presents, for the first time as a collection, some articles from the *American Fastener Journal*. Appendix E adds some concluding thoughts and application topics. I have learned that some tasks come more easily, some are more difficult, if not all. I realize that even the most accomplished writers make errors, which can survive through to the final published work. I have further learned that I am neither particularly accomplished in writing, nor adept at avoiding, editing out, and otherwise providing written proof to the reader of my lack of error-free composition. All such errors occurring in this edition are all my own. I request your sympathetic stance when finding all such lapses and request you to accept them, knowing that I am truly contrite.

Preface to the First Edition

Mechanical Fastening, Joining, and Assembly is intended as a guide and reference source for product designers, engineers, manufacturers, and students interested in a solid foundation in assembly and joining. A product is only as strong as its weakest link. For many products, the weakest area is at the joints where the components are assembled. Well-fastened assemblies result from good design, quality parts, and properly executed assembly procedures and production processes. Personnel having these responsibilities benefit from knowledge of mechanical assembly engineering and fastening technology. Observing the current generation of product designers and engineers, one cannot help but be awed by their grasp of electronics and computer-based designs and all of the associated skills. But also observable is a decline in attention directed toward mechanical technology as the newer electronics-based products gain favor.

This, in itself, is not cause for concern so long as the mechanical products and processes perform well. Mechanical proficiency, however, can decline with time unless studied and applied. With prolonged inattention to the fundamental mechanical technologies, including fastening and assembly work, progress in these other areas will show the detrimental effects and not advance at a pace and intensity that will do justice to the products that these newer technologies make possible.

Chapter 1 explains the fastening function in depth. In Chapters 2 through 4, you will learn the types of fastening approaches that can be used together with examples of what they cost and how they work. In Chapters 5 through 9, specific joining applications, including vibration, standard and special materials, and environmental factors are detailed, and useful reference charts are included for future use. The appendices serve as a useful reference for fastening terms and concepts and give examples of the current generation of high-efficiency assemblies and fastening/joining trends for the future.

Fastener companies, the technical community and its engineering societies, standards organizations, and academia all provide useful and valuable engineering knowledge on fasteners and fastening. But by their nature, they are more narrowly focused and/or constrained in scope by time and economic considerations. This book's goal is to help bridge that gap.

Current and future generations of product designers, engineers, manufacturers, and students should have access to a solid knowledge base of mechanical fastening, joining, and assembly information. The output of their creative efforts should yield the highest integrity, efficiency, and analytical ability of fastening and joining.

Mechanical Fastening, Joining, and Assembly will enable readers with a starting interest in assembly technology as well as those with a more in-depth background in this field to gain a deeper understanding of fastener and fastening design and point the way to assembling more efficient and competitive products.

Acknowledgments

As I grow older, I increasingly realize the important roles that many of the people I have been fortunate to have had in my live have fulfilled, many with great distinction. First, mom and dad. We always had lots of books and reading materials, and they set a good example by reading them. An early steam engine I was given as a gift, I am sure at my dad's behest, was a favored toy that I learned to operate, take apart to understand its operation, and then continually tinker with to make changes and what, to my childish eyes, were improvements. Second, my mom's three brothers, my uncles, who each in their own way taught me important lessons about airplanes and jet engines, wood as a raw material and as a finished product, the requisite skills of making things work, and the marvels of electricity and electronics. More importantly, I learned by their good examples, how a man can hone his individual craft and make his way successfully in this world. If I pass along any lessons here, they would be "there is no such thing as can't," "a strong back helps but a strong mind is the key tool in your personal tool kit," and "if you have a good mind and good hands, you'll never go hungry." Thanks guys, I hope I have "done good." Many thanks to Mike McGuire, publisher/editor of *American Fastener Journal*, among many other accomplishments, both encouraged and motivated me to write when he started his magazine. When I went looking for that graphics upgrade, he recommended Cyndi Daines who on this project was the epitome of both professional and clutch. To all of the people I have been honored to work with, from my first year as an apprentice, through my most recent position as a regional manager for the Johnson Gage Company, you have my deepest gratitude.

A word of deep thanks is due to the good people of Taylor & Francis, starting with Jonathan Plant who got the project rolling, to Amber Donley who kept me "on the rails" during the start-up and well into the project, and to Jill Jurgensen who "came in during the late innings" when needed.

Last, at her acknowledged spot at the top of my list, my wife of over 30 years, Lillian, who was a rock-steady presence by my side during the first edition, along with our four wonderful children, Sarah, James, Holly, and Jason. She is now with us in spirit for this second edition. Lill always said I liked to do everything twice. So, "Hey you, here's two … until we once more can say again once more, 'Siamo, Tutti Qua.'"

Author

James A. Speck, PE, worked as a regional manager of the Johnson Gage Company, Bloomfield, Connecticut, following an extensive career in the fastener industry. He has trained and served as a technical expert in the Department of Commerce's National Voluntary Laboratory Accreditation program for fasteners and metals. Jim has presented numerous workshops and seminars on fastener fundamentals, fastener applications engineering, and screw thread technology. Jim is also the author of numerous key publications. After completing an apprenticeship at the Philadelphia Naval Shipyard, Jim went on to receive BS (1971) in business administration from Drexel University, Philadelphia, Pennsylvania, and an MS (1974) in management from the Hartford Graduate Center, Hartford, Connecticut, and he is a licensed professional engineer in his home state of Connecticut.

1

Fastener Functions and Assembly Testing

1.1 The Fastener Design Role

Fasteners are used in applications for strength, appearance, and reusability. Considering them in reverse order, reusability is often a consideration since many assemblies may need to be taken apart for maintenance, service, and repair. Even in some moderate-volume assemblies, it may be desirable by either some percentage of the target market or the product manufacturer, if not both, to have the ability at some time after manufacturing assembly to disassemble. Even for those products that have only a remote possibility of ever needing disassembly for service or repair, there is no way of knowing which of them will need to be disassembled. If only one in a thousand products needs reusability in its fasteners, for whatever purpose, the reusability of the fastenings is a design requirement. Since knowing which 1% will require reusability is not possible during assembly, it is a requirement for all of the assemblies produced during the production run.

Fasteners play an appearance role in the market acceptance (or lack of acceptance) of their respective products. Although, when asked, many purchasers will state that economic or technical features of a product were a prime decision factor in purchasing, it cannot be refuted that style and appearance are much larger factors in what we choose to buy than many of us would care to admit.

Fastener appearance can function as a visible sign of robustness as in the exposed, high-strength fasteners in a piece of industrial machinery. In this manner, they can give visual assurance to customers and prospects of product strength, durability, and value. Conversely, fasteners can be designed "out of view," a design trend that is popular now in consumer durable goods, especially automobiles and appliances. This well-tailored, sleek appearance makes a strong visual statement of thoughtful design, seamless assembly, and a general up-to-date, modern appearance. These two and numerous other fastener "looks" ebb and flow in popularity. But the assembly designer and manufacturer ignore them at their risk.

At the most basic level, fasteners are used because of their strength, especially their holding strength in the assembly. As fasteners come "out of the box," they make available potential holding power. Holding power is the key element of fastener strength. Combined with appearance and reusability, holding ability is the mechanical foundation of all fasteners.

A simple example can emphasize this point. Let us say that we have an assembly composed of some flat rectangular parts, namely standard playing cards. Figure 1.1 shows a sketch of them, with some simple forces applied.

If we have an assembly goal of holding them flat together, we have several options. We could simply hold them by hand. This would certainly work but would be inconvenient, not to mention impractical, for any length of time. The original carton would keep them loosely together and in one place. Being paperboard though, it may not give many reuses. At that point, we would be faced with making another fastening decision.

A logical choice is an elastic band. It meets, in some measure, our three fastener criteria of strength, appearance, and reusability. Our rubber band will certainly have the strength to hold the cards together if we select a band of a suitable size. It resists loads in the x-, y-, and z-directions. In the x-direction, friction between the cards multiplied by the clamping force of our elastic band fastener resists shear loads that would allow the cards to slide apart. We could estimate the shear strength by the following equation:

$$Shear = Tension \times Friction$$

In the y-direction, the elastic band fastener resists tensile forces tending to pull the cards apart by applying a counteracting compressive force. And in the general z-direction, our fastener winds up to apply resisting torque to oppose any twisting motion applied to our "assembly." So our strength fastening conditions have been met. True, we have not tested or qualified our fastener's strength. And even though we have satisfied the static tension, shear, and torsional service loads, we know little about the service life of this assembly or its response to dynamic loading. Dynamic loads in this simple

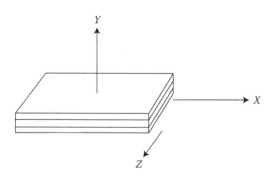

FIGURE 1.1
Cards with forces.

example, such as those resulting from a dropping of the card deck, would constitute a reasonable design consideration.

What about reusability? We have certainly satisfied this goal. We can install our fastener and also remove it many times. We can go through assemble and disassemble cycles repeatedly. No special tools are required, and our fastener is easy to access. We have scored well on reusability. We may lose some elasticity with time or ultraviolet light exposure. We could address this if they are service application factors.

How did we do regarding appearance? We have some latitude here. If we choose a utility quality rubber band fastener, it may not look particularly well assembled. But we could opt to carefully orient our application of the band as we put it on and choose one that only requires one wrap to hold fast. This would also increase assembly efficiency. We can even select a color to match the cards.

Obviously, this simple playing card and rubber band fastener example is oversimplified and far removed from most industrial and commercial fastening applications. Our hypothetical card assembly problem is useful in that it lets us look at strength, appearance, and reusability as factors to consider when designing and installing fasteners. Most of the assembly applications we fasten are more demanding.

1.2 Analysis of a Simple Metal Assembly

As an example, let us reduce the number of flat rectangular members to 3 and make them all out of steel. We have several factors to consider in how to fasten our assembly. The first would be reusability. If we consider the assembly's degree of serviceability as falling in a range, we could position it somewhere along a linear function. Figure 1.2 shows this graphically.

If our assembly never needs to be disassembled, some permanent fastening such as welding or adhesives may represent our best fastening option provided strength and appearance criteria are satisfied.

If more disassembly is needed, unthreaded fasteners such as clips, snaps, pins, or a press fit fastener might be our best option. Within this class of fastener, strength, purchasing factors, and appearance would further guide our decision regarding fastener type.

Never serviced   Frequently serviced

FIGURE 1.2
Likert scale of assembly service needs.

Going up the scale of disassembly, threaded fasteners along the lines of bolts, nuts, screws and tapped holes, self-threading screws, and self-drilling and self-tapping screws would be possible options, after which strength, purchasing factors, and assembly appearance factors can be factored into the decision of which fastener type to use.

At maximum reusability limits, quick acting fasteners such as quarter-turn fasteners, captive fastener assemblies, and similar fasteners are options.

Anywhere along the reusability range, from "never needs to be disassembled" to "frequently taken apart," holding strength becomes the next significant factor. To determine the amount of fastener strength, or holding ability, to design into our assembly, we need to know the materials our assembly is constructed of. We also need to know the directions and sizes of service loads to which it will be subjected.

For our consideration here, we will assume that the assembly plates are made from AISI 1010 steel, are not heat-treated, have a yield strength of 50,000 PSI, and a tensile strength of 85,000 PSI. They are 1/2 in. thick each and are 4 in. × 8 in. in area. Figure 1.3 shows a sketch of our plates.

Let us say that while in use by the user of this assembly, they must withstand a tensile load or pull of 10,000 pounds, perpendicular to the plates. We must have a fastener, or fasteners, which at the least resist the 10,000 pound tensile load. As a starting point, we can locate one tension fastener at the center of gravity, or centroid, of the plate plan area. We will make it of material and dimensions with the potential to resist the 10 K service load with a reaction of this magnitude or greater. Of course we have not precisely located the service loads. And several other factors require our attention. Principal among these factors is the best number of fasteners for our application. We have initially selected one. Is it enough, or are additional fasteners needed? Further, how do we size the fastener(s)? And how do we provide for its/their anchoring in the assembly? Also, how will we accomplish an effective installation of our fastening?

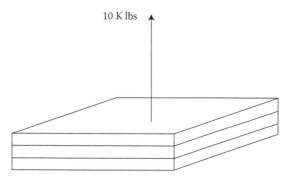

10 K lbs

FIGURE 1.3
Tensile service load.

Let us look at the number of fasteners to use. In the application, we appear to have used reasonable mechanical judgment. One of our centrally located tension fasteners has a center of action that is only 2 and 4 in., respectively, from the edges of the clamped members. But without accurate information about the placement of the 10,000 pound service load and the assembly member's reactions under service load conditions, we do not have sufficient information to reach an informed conclusion about the efficient number of fasteners required for optimal fastening.

Why not place one in each corner? Or one in each corner plus a center fastener for a total of five? This may not improve our assembly. First, adequate room for the fasteners means less room in the assembly for more functional, value-adding working room. The additional stresses from four or five fasteners on the assembly members could cause unwanted distortions in the assembly. And equally important is the added cost of the additional fasteners raises the cost of assembling our product. And the fasteners, by themselves, represent only a fraction of the cost of the fasteners installed in the plates. Depending on the type of fastening selected and the method of anchoring it, the installation costs can represent a four times of the fastening purchasing cost. Put it another way, the total cost for a fastening can be made up of 20% fastener purchasing costs plus 80% fastener installation costs.

$$\text{Total fastening costs} = 20\% \text{ Fastener purchasing cost}$$
$$+ 80\% \text{ Fastener installation cost}$$

The purchasing costs represent the fasteners, their freight, handling, and disposal of packaging and material handling. Significant costs are also associated with the costs of physically handling the fasteners. Installation costs are made up of the work performed to prepare their site, work done to put them in the site, and costs associated with any tooling and fixtures used. It is very apparent from this that these can represent significant costs.

Clearly, there is a compelling motivation for any manufacturer to control these costs. Fastening efficiency requires the reduction of these costs to levels appropriate for the safety and structural integrity of the application.

A reduction in fastening cost at too great a rate could compromise safety and put both the consuming public and the assembly manufacturer at risk. An optimum fastening level is required. Graphically, an efficiency function curve is a good way of viewing fastening costs. Figure 1.4 shows fastening efficiency with respect to the number of fasteners used. It plots fastener installation numbers and costs as functions, with fastening efficiency representing an ideal range for both consumer safety and producer mechanical advantage.

Returning to our application, let us view the 10,000 pound service load as being composed of two 5000 pound pulls tending to separate out three plates. As assembly designers, not only do we want to prevent plate separation, we also want to avoid the plate edges from lifting up. Good mechanical construction would be to position two fastening sites, each located close to

Number of fasteners used vs. efficiency

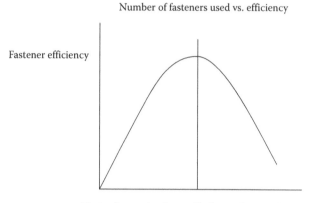

Fastener efficiency

Under-fastened   Optimally fastened   Over-fastened

FIGURE 1.4
Fastener efficiency graph.

one of the service load points of action. The fastener clamp loads balance the service loads, preventing both plate separation and plate edge bending. Our assembly is in a condition of stable equilibrium.

But what if our confidence in the 10 K service load is less than 100%? Good mechanical judgment would call for us to provide additional clamping load to provide for any over-10 K pound service load peaks or spikes. A standard approach would be to provide for a design safety factor. This multiple will allow for unforeseen service load application to our assembly, with no assembly failure in the form of plate separation. Design safety factors represent a form of mechanical insurance. The size of the design factor, or insurance, should be chosen to be appropriate for the nature of the assembly and its intended market.

If our market can be safely judged as very low risk in nature, a design factor of 2 might be appropriate. With more unknown risk, a design factor of 10 is conceivable in applications where human safety and health are involved. An example would be where catastrophic failure would result from assembly failure. Redundant fastening in addition to the high design safety factor is appropriate in these cases. Our assembly would be fastened at different levels depending on the design factor decisions made. Figure 1.5 shows a model for this fastening decision.

The selection of fastener type and placement requires detailed knowledge of the size of any possible service loads, their direction and location, as well as consideration of reusability, appearance, and economic efficiency.

Once these decisions are thoroughly evaluated, appropriate design factor selection requires full consideration of the risk of assembly failure. For our case, let us say we select a design factor of 5 after carefully investigating the service factor risks. Our next decisions involve selecting the style of tension fasteners, sizing them, and deciding on an anchoring method in the two sites we have chosen. For style, let us choose hex head cap screws as our fastener

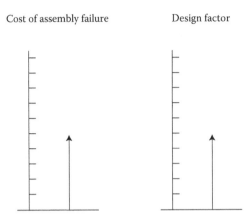

FIGURE 1.5
Failure to design factor comparison.

type. They are readily available. Tightening tools are common and fit the industrial style of our metal plates. Since they have machine screw threads, they are very reusable. An interesting observation is the nomenclature difference between the bolts and screws. Although they are commonly used interchangeably, a bolt is used with a nut, and a screw, as normally used, is threaded directly into the assembly.

Since we want to keep our parts count down, we will use two machine screws. There are some technical advantages to using washers and nuts in this application—one among them is a longer grip length. This gives us longer total elongation for a given preload. It also means that we will be applying the clamp load over a larger bearing area giving a greater clamp-affected zone. This will give us lower bearing stresses. This is important if the fasteners are significantly harder than the assembly. We will forgo these advantages to reduce the parts count but will not lose sight of the trade-offs we have made to reduce fastener count. In an application, we would expect to thoroughly test our assembly using both laboratory and field work, possibly combined with computer simulation. Figure 1.6 shows other possible loads.

With a design safety factor of 5, a service load of 10,000 pounds, and two screws, each screw will need to be capable of carrying 25,000 pounds. This presupposes uniform loading. In actual practice, uniform loading is not always the case and in designing a fastening system, a nonuniform loading, and possible resulting bending stresses, should be added to our calculations.

To determine static fastener strength, we would need to know the fastener strength of material and cross-sectional area. To determine the critical area of our fasteners, we can rely on industry standards, which have been developed to help size cap screws for their intended service load. For our purposes, we will specify an SAE grade 8 material with 150,000 psi (lb/ft^2) of tensile strength and 130,000 lb/in^2 of yield strength. Our screws will be 1 1/2 in. long, and we

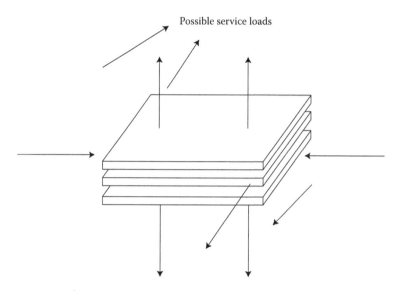

FIGURE 1.6
Loading on three-plate example.

will specify that the threads be the full length of the shank. We can start setting up equations to find the required cross-sectional area as follows:

$$\text{Target clamp load} = 25,000 \text{ lb}$$

Select a screw type that is thread critical. A thread-critical fastener is one in which, at ultimate strength, the threads are, by design, the area of fracture. Figure 1.7 shows the most highly stressed areas of the hex head cap screws we have selected.

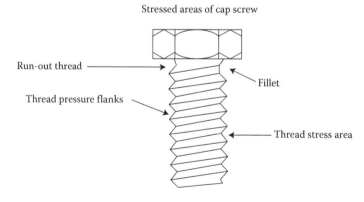

FIGURE 1.7
Highly stressed areas of screw.

Calculate the needed thread area:

Stress = Load/Area
Stress × Area = Load

For stress, we must select an allowable loading rate. We will use a loading rate of 75% of the fastener's yield strength of 130,000 lb/in.^{2}.

$$130,000 \text{ lb/in.}^2 \times 0.75 = 97,500 \text{ lb/in.}^2$$

$$25,000 \text{ lb/in.}^2/97,500 \text{ lb/in.}^2 = 0.2564 \text{ in.}^2$$

We can convert this into an equivalent diameter, which will indicate the size of fasteners we will be using:

$$0.2564 \text{ in.}^2 = \frac{3.1416 \, d^2}{4}$$
$$= 0.7854 \, d^2$$

$$0.2564 \text{ in.}^2/0.7854 = d^2$$

$$\text{Sq root of } 0.2564 \text{ in}^2/0.7854 = d$$

$$0.571 \text{ in.} = d, \text{ our screw diameter}$$

A thread larger than 1/2 an inch in diameter and smaller than 3/4 will work. If we look at just the thread section, we can see that the thread stress area is offset by the angle that the threads are inclined. The thread stress area is an area around the pitch diameter, partway between the thread major and minor diameters.

From the following charts for inch and metric threads, we can see that the 5/8-11 screw will be a good size for the load we have specified and the material we plan to use. Charts such as these for every type of fastener that a designer may consider are useful in selecting the size of a fastener for a given application and material conditions that will adequately handle the intended loads with the desired design factor while avoiding either oversizing or undersizing the fasteners. Fastener suppliers can be useful sources for charts on their fasteners.

Inch Series Screws (Coarse Series)

Nominal Size (in.)	Stress Area (per in.^{2})	Threads (in.)
1/4	0.0318	20
3/8	0.0775	16
1/2	0.1419	13
5/8	0.226	11
3/4	0.334	10
7/8	0.462	9
1	0.606	8

Metric Series Screws (Preferred Pitch)

Nominal Size	Stress Area (mm²)	Pitch
M5	14.20	0.8
M8	36.60	1.25
M10	58.00	1.50
M12	84.30	1.75
M16	157.00	2.0
M20	245.00	2.5
M24	353.00	3.0

We made a choice here regarding the threads per inch, or thread pitch if using the metric system. Both coarse and fine thread pitch series have benefits and drawbacks. A comparison shows the following facts.

Thread Pitch Selection

	Coarse Pitch	Fine Pitch
Benefits	Fewer turns to tighten	Larger stress area
	Easier to start	Slightly higher vibration resistance
Drawbacks	Prone to loosening (from Reference 1)	Prone to hard starts

Using a very basic example, we have set a fastener design goal, quantified the service loads, selected a suitable design factor, and sized the fasteners. To complete the process, we should perform some application testing to assure and document our assembly design. If the assembly operation is to be repetitious, assembly materials and procedures should also be documented.

In testing, our goal should be to subject our assembly to test loads as similar as possible to those it will work under in service. Since in this application, a 10 K tensile load is expected, composed of two 5 K components, we can use this loading for baseline testing.

A tensile tester and some fixturing will enable us to perform these qualifying tests. In actual practice, a test setup similar to the following illustration would be found. The tensile tester would gradually apply the load until 10 K pounds are acting to separate the plates and being resisted by the installed fasteners. Figure 1.8 shows a fastener tensile test setup.

Another useful testing procedure would be to inspect the test-loaded assembly, with indicating gages such as dial indicators for measurement of plate movement after assuring that the assembly is secure.

During this initial testing phase, a statistically significant number of tests would be performed to give to the assembly builder confidence in the mean and range values of the test results.

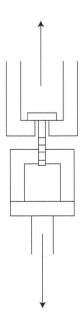

FIGURE 1.8
Fastener tensile test setup.

Assembly testing to ultimate strength can also provide the assembly designer with useful data on the performance of the assembly, provided proper safety precautions are taken. Suitable surrounding enclosures should be in place, which can withstand the impact of fastener and assembly pieces being propelled at them with the force of the released energy. Consider that the stored force in two 5 mm nominal diameter and heat-treated steel machine screws could equal many thousand pounds of force. At ultimate assembly failure, this force is transferred to the fractured assembly components and fasteners. Sometimes, the resulting accelerations can be dangerously high if the fragments are not safely contained. Examination of the failure mode can provide very useful information. In addition to giving validation of the safety design factor built into the assembly, it shows where the critically stressed areas of the assembly are located. This, along with the ultimate failure load data, and the computed failure area stresses, can be extremely helpful in revising these areas. Revisions to the assembly/fastening design can include changes for increased robustness prior to production or product launch, if needed.

Another test we might consider is to apply a sudden, large test load rather than the more gradual one of the tensile tester. An impulsive load is defined as one which is applied in a time less than the natural frequency of the assembly. The assembly natural frequency would be a function of its assembled mass, found from its weight divided by a gravitational constant, its relative spring rates, and the assembly's degrees of freedom. Our application has three steel plates with a weight of 4.5 lb each and a total of 13.5 lb. Our two hex head cap screws together weigh 0.2 lb. Our total assembly weight is 13.7 lb. If our natural frequency were about 0.7 cycles per second, we would want to be alert to assembly loading cycles around this frequency (2).

As a practical consideration, a commercial laboratory would be a good choice for performing impact-type work on the relatively small test iteration scale that a new assembly would entail.

With all of the test equipment used in assembly qualification and design evaluation, test procedures should be documented and the appropriate training should be in place. Measurement and test equipment should be on a calibration schedule with comparisons made to know standards traceable to an established source, such as the National Institute of Standards and Technology (NIST) in Gaithersburg, Maryland. And test results should be recorded in a professional manner, and appropriate record maintenance kept for future reference.

If possible service for the assembly includes temperature-extreme environments or extended use in corrosive atmospheres, tests under these conditions are advisable. Laboratories for performing hot and cold tests, along with salt spray and humidity cabinet, or other corrosive atmosphere conditions exposure testing are available.

Other tests could include ones for vibration in several different modes to document the assembly's ability to maintain a significant percentage of fastening force under the rugged conditions found in some applications.

If the service load is an alternating, defined-frequency-range environment, typical of the loading conditions found in many prime movers such as internal combustion engines, fatigue life may be an important fastening design consideration. In alternating load applications, control of fastener preload, joint relaxation, and joint face embedment should be primary engineering factors.

Tests that we have conducted would include fatigue testing. In a fastener fatigue test, an alternating load is applied about a preestablished mean load, which is preloaded before starting the test. Using actual production assembly components and assembly fasteners in the fatigue test assures the closest replication to real-world service conditions.

However, using just the fasteners in test fixtures can be a valid fatigue testing procedure for periodic inspection and candidate vendor evaluation, or for competitive fastener design testing.

The fatigue tests are conducted to generate pairs of alternating service load and service load cycles. An advanced form of the fatigue test relies on driving the test specimen into resonant frequency. This extreme form of alternate loading results in cascading loads at resonance, thereby reducing the time to complete testing.

1.3 Fastening a More Complex Assembly

Many components that we are required to assemble present joining surfaces and service loads that are not as regular in either their geometry or engineering mechanics. The fastening, joining, and assembly procedure decision to be made still must address the basic issues of assembled strength, assembly reusability, and assembled component appearance that form the basis for our fastening decisions in the prior, albeit simplified, example.

Let us consider an assembly comprised of two injection-molded plastic shells clamping a die-cast, light-metal pump housing. The housing contains a gerotor pump mechanism and is designed to be mounted onto a fractional horsepower induction motor. Our pump assembly is to be designed and marketed to industrial, agricultural, and the home do-it-yourself market. The following illustration and specification list give the application specifics, less fasteners. As an example, it is intended to show some of the data needed and analysis that is useful in carrying out the planning and installation of a successful fastening application. Time spent in this type of fastener analysis can help avoid costly fastener failures later in the design and manufacturing cycle.

We will review the application and go through the steps required to choose a fastening approach/size and install our fastener(s). We will conclude with a review of our assembly and tests we will conduct, and we will suggest possible directions for future assemblies. Figure 1.9 shows our assembly.

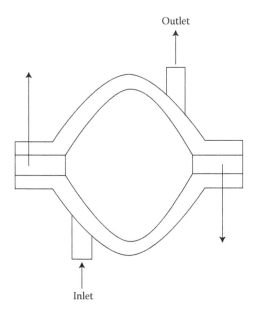

FIGURE 1.9
Pump assembly with forces shown.

Pump Assembly Specification

Materials
Shell Halves ABS Flexural Modulus 140×10^3 psi Pump Housing Zinc Die Cast Y.S 35×10^6 psi Gerotor Corrosion Resistant Stainless Steel Y.S. 40×10^6 psi
Grip length 0.500 in.
Maximum internal working pressure 120 psi
Temperature range –10 to 110°F
Distance from centerline to edge 0.25 in.
Serviceability—not field repairable

Let us take the factor of reusability first in our evaluation of the assembly for fastener selection. Field servicing is not a factor. In this situation, we do not have to necessarily rely on machine screw threads and tapped holes to facilitate disassembly. This opens up some additional options for our fastening approach.

It is possible that we may want to disassemble at some point after the useful service life is expended. Two possible situations would exist if it was desirable to recover the pump materials for recycling.

It is also possible that we may want to repair units back at the factory either as an after-sale service or as a method of evaluating part wear and performance life.

We have several options open to us. Self-tapping screws could be used to eliminate an internal threading operation. We could also use either solid or semi-tubular rivets to clamp the pump components together. Another candidate would be blind rivets. Of these four fastening options: self-tapping screws, solid rivets, semi-tubular rivets, and blind rivets, we can lay out the advantages and drawbacks of each fastening approach.

We could make a comparison of these four choices by ranking their relative strength in the areas of our interest:

Self-Tapping Screws	Solid Rivets	S.T. Rivets	Blind	Rivets
Degree of reusability	High	Low	Low	Low
Strength/size ratio	High	High	High	Low
Purchasing cost	High	Low	Low	High
Installation cost	High	Low	Low	Low

As an additional note, we could also evaluate chemical adhesive joining methods as well as mechanical snap/press fits or staking of the parts together. We will forgo these choices in our evaluation at this time. If sized properly, all four of the choices will deliver the needed clamp load to hold our assembly together. An important selection factor in this or any production assembly application is the number of assemblies that we plan to produce.

Realizing that exact prediction of future assembly build rates may be very difficult, it is still important to establish an estimate of the assembly build quantity. Build quantity will set the level of installation systems costs—the effort and the tooling, machinery, documentation, training, and inventory needed to get the assembly work done. And it is these costs that can represent, as indicated earlier in this chapter, the major percentage of the assembly cost. Equally important is it fixes some of these costs for a long time into the assembly's commercial future, if not for its entire life.

After evaluating our four choices, we will choose a semitubular rivet as best meeting our assembly's design goals. Its strength can be sized and tested to deliver the holding power that our pump will require during its service life. While designed to be permanently joined in the field, semitubular rivets can be easily drilled out for some after-service work such as recycling, wear evaluation, or factory-authorized repair.

The appearance of our two assembled joints in the pump can be made of a material and plated or finished to give a range of esthetic looks as we require. By selecting a head style and riveting technique, we can match the semitubular rivet's shape to the pump housing's geometry.

Sizing the rivet is our next step, which requires us to know the pump service loads that the rivets will be called on to oppose for the life of the assembly. From our data, we know that the internal pump pressure is 120 psi. Let us say that by calculating the internal area of the pump housing, we arrive at an internal working area of 7.5 in². From the pressure and area, we estimate the primary service load as

$$\text{Stress} \times \text{Area} = \text{Service load}$$
$$120\,\text{lb/in}^2 \times 7.5\,\text{in}^2 = 900\,\text{lb}$$

We will need to clamp the assembly with a compressive force greater than 900 lb. In actual practice, we may want to use a pressure gage to confirm this internal pressure. Testing service loads is often good practice. It can often more than justify the setup and testing expenses involved. In some cases, loads not foreseen during assembly design can be brought to the designer's attention. In a specific example, let us plan on a shear load from the internal torque of the gerotor pump mechanism. An accurate assessment of assembly service loads is a key factor in the efficient sizing of the fasteners specified. By knowing both the normal and extraordinary service loads of an assembly, coupled with a suitable design factor, we can have confidence in the strength, durability, and integrity of our fasteners. Often, prototype testing and computer simulation can be used to increase our understanding of service loading amplitudes, directions, and frequencies. Let us take a look at the rivets as they will be installed in the pump and at load. Figure 1.10 shows the assembly.

We will specify a semi-tubular rivet which has a shank diameter that can develop and hold a design factor multiple of 450 lb for each rivet. It will also be specified that the shank length is sufficient to grip our assembled stack of components with sufficient rivet body to rivet over into an acceptable joint.

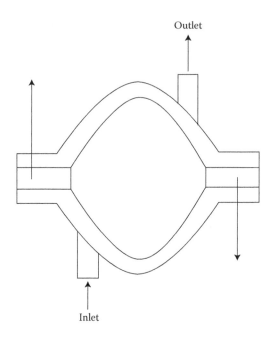

FIGURE 1.10
Pump with rivets assembled.

We will also say that the pump action from the internal workings of our pump exerts a torque, which in turn results in a shear load of 100 lb on the pump housings. Summarizing our service loads with an equal share to each rivet, we have a tensile load of 225 lb and a shear load of 50 lb on each rivet. Making a force diagram of the load that our rivet will carry would give us a layout such as the one illustrated in Figure 1.11.

This gives us a resultant load of 231 lb at an angle of approximately 77.5°. One step we can take is to design a fastening feature into the pump housing's halves to support some of this shear load. A simple tab-and-socket design into the injection molding dies and die casting tooling will transfer some of the shear load to these formed-in features, thereby making the assembly part of the fastening system. When working with a new design, it is also helpful to take a systems viewpoint in analyzing how and where to fasten. Instead of just looking at a particular fastener to hold the service loads, consider the fasteners and assembled components, working together to accomplish this goal. A simple tab-and-socket would function like the sketch in Figure 1.12.

Based on our application information, we can select a rivet material from any material that is formidable. For our case, we will use 300 series stainless for its corrosion resistance and also its relative strength, with a yield strength of 35,000 psi and a tensile strength of 85,000 psi. We will also select a design factor of 2.

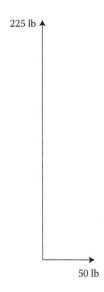

225 lb

50 lb

FIGURE 1.11
Pump rivet forces.

We set up the following equation to select a diameter:

$$Area = Load/strength$$

$$Area = 462 \ lb/35,000 \ psi$$

$$Area = 0.0132 \ in.$$

Pin one piece of pump housing

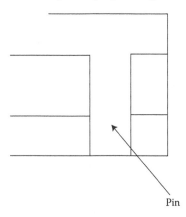

Pin

FIGURE 1.12
Pin one piece of upper pump housing.

Solving for shank diameter, we obtain:

$$\text{Area} = 0.7854 \ \text{diameter}^2$$

$$0.0132 \ \text{in.}^2 = 0.7854 \ \text{diameter}^2$$

$$0.0132 \ \text{in.}^2 / 0.7854 = \text{diameter}^2$$

$$\sqrt{0.0132 \ \text{in.}^2 / 0.7854} = \text{diameter}^2$$

$$0.1296 \ \text{inches} = \text{diameter}$$

A shank diameter of 0.130 in. will carry our load. At this point, we can make one of two specifying decisions. If our applications were either very unique in some regard, or of a much higher volume build rate, having a rivet custom manufactured to our unique specifications may be the most economical and efficient design approach. Since that is not our design situation, it is probably in our best interest to select a standard rivet dimension, which best fits our requirements.

For this work, we can turn to the technical standards. For this particular class of mechanical fastener, the American Society of Mechanical Engineers (ASME), along with the Industrial Fastener Institute (IFI), publish standards that will help in our specification work. The applicable specification for semi-tubular rivets is American National Standards Institute (ANSI)/American Society of Mechanical Engineers (ASME) B18.7. Standards published by ANSI/ASME cover a wide range of industrial fasteners and provide designers and manufacturers with useful guidelines for fastener selection. Consulting with ANSI B18.7, we can see that a 0.146 diameter standard rivet is the next larger diameter over our calculated diameter and will give us a margin of strength over our requirements.

We also must choose a shank length. Since our gripped length is 0.500 in., we select a fraction inch length of 0.625 in., which gives us 0.125 in. of rivet material to form the clinched portion of our installed rivet. In using rivets of either the tubular, semi-tubular, or solid design, be sure to allow this "riveted" shank material in your fastener and assembly plans.

We next have to select a head style. If we refer to the standard we can see that oval head, truss head, and 150° flat countersunk head styles are standard. It is important to note here that although a configuration is published as a standard, it may not necessarily be easy to obtain. The designation standard only signifies that a published specification, or "standard," exists for the design. Knowing whether it is a "stock" item is in this case more important. For this reason, involving the fastener supplier early on can give guidance on availability, which can make later procurement easier.

Since we will have a strength gradient between the stainless steel of our rivet and the engineering grade plastic of our pump housing, we will choose the truss head, which has a larger bearing area or "footprint," and will distribute its bearing lead over a greater area. This will reduce the compressive stress on our plastic components.

Next we will need some tools to rivet our assembly together. We can again consult fastener and industrial distributors, rivet suppliers, and industrial directories for suppliers. For our purposes, we will select an orbital riveter with a force capacity, speed, and economics that will suit our build requirements, with perhaps a reserve capacity to accomplish increased production speed to meet future needs without having to obtain another orbital riveter.

Our next steps will be to assemble some trial pumps and conduct both proof and ultimate strength testing as we performed in the previous example, as well as any other mechanical or environmental test appropriate for our intended service. Lastly, we would want to completely document our design steps and data, assembly and tooling steps and procedures, as well as the necessary inspection and testing requirements. As can be seen by these examples, good assembly design practice calls for following specific technical steps combined with accurately established assembly engineering requirements. In the next chapter, we will see an expanded range of fastening options.

1.4 Fastening Real-World Applications

We will now turn our attention to a threaded connected assembly using dissimilar materials. Fastening dissimilar materials is an often-encountered assembly application. It presents specific application factors, which must be planned by ever stakeholder: from the manufacturer and designer, through the supply chain members, maintainers, using military jargon for this function, to the end users and recyclers.

Let us consider a rotating aerospace assembly composed of a large-diameter aluminum hub, a steel-inserted composite material blade assembly, a set of screw thread inserted into internally thread aluminum hub threads and bolted together with 12-point drive high-strength steel bolts. Our service factors present the fastener designer and user with some imposing performance requirements.

As this is a threaded fastening, let us discuss the reusability of this assembly. In the screw thread section of the propeller assembly, the external bolt threads are during assembly with the internal threads of the internal insert screw threads. This clamping load is applied and supported by the external insert screw threads on the internal screw threads of the aluminum propeller hub. The geometry of the screw thread is composed of the angles, lead and circularity features of the screw thread flanks, lead, pitch, and roundness of the major, minor, and pitch cylinders. The bolt, insert, and hub screw thread geometries must start and remain in their proper geometric limits if the screw threads are to allow good initial assembly and as importantly, subsequent removal and reassembly of the propeller during the engine and aircraft's service life. The reusability will be required to adequately carry significant static and dynamic loads if reusability is to provide performance that all involved can consider as good reusability performance. What reusability performance measurement can be employed to measure this assembly's reusability?

This is an important question. From a practical standpoint, most people would use a simple "If it works" reusability measurement criteria. If we are able to install, remove, and reinstall the propeller by screwing and unscrewing our bolts, we have a successful design. And for many users and applications, this is satisfactory. That said, it does not enable us to quantify our reusability. It also does not provide sufficient guidance for measuring the degree of reusability: identifying existing or potential weaknesses in the assembly's reusability. For continual improvement of reusability, with respect to both the total number of cycles available or the cost–benefits associated with fastening alternatives, numerical metrics are needed. Let us consider some of these.

Since we will be applying torque to the 12-point bolts to both install and remove the propeller, an index of torques might prove useful. One used in industry has been the prevailing on and prevailing off torques. For example, if we require 100 N m of tightening torque applied to the wrenching surfaces of the 12-point head torque to tighten the screw to its design preload target, the 100 N m can be the denominator of our index. We can record the loosening torque as the numerator of this index. In tightening the bolts, clamping load is being developed in our assembly by bolt elongation, resisted by propeller compression. When the tightening torque is applied, work is being performed on our assembly "system". A good analogy is pushing a block up an inclined plane. The inclined plane angle is the lead angle of the screw threads. If our bolt has M12 × 1 6 g screw threads, our lead is 1 mm and our lead angle is 1°, 42 min, 13 s.

During untightening, the block is coming back down the inclined plane, work is being performed by the system and the loosening torque is generally lower. For this example, let us say 85 N m giving a reusability index of loosening and tightening torques of 85/100 N m, or 85%. A range of this reusability index would provide a quantitative measure for comparison and process improvement. If subsequent lots have increasing loosening torques, our system's reusability has decreased. If we develop a new source whose components assemble with 100 N m tightening torque but loosen with only 75 N m, then we have changed the efficiency of our reusability. A customer need not apply as much loosening torque to remove the propeller. While not seeming to be significant, one assembly cycle, over the life of the system, or more to the point, over the life of a complete propeller assembly production run of large numbers and years of service, these efficiencies can add up to significant savings in maintainer effort, energy usage such as for compressed air, and electrical power and tool life. As an example in the opposite direction, a screw-threaded component subject to galling in the threads may actually "lock in" and require a higher torque to untightening, making this much less efficient.

Building on this index, let us add the factor of time. On the tightening side, we could compare quantitatively the initial tightening torque of 100 N m with the second, third, fourth, fifth, and so on. We might find decreases of 2%, perhaps more or less. Factors such as contacting surface smoothing, minor local permanent deformations from material yield, and other assembly component changes could be evaluated and understood. This phenomenon was often referred to as "bedding-in" of an assembly. We could also plot with torque and time graphical axis the shape and duration of the tightening torque. This would give us a tightening torque signature.

On the loosening side, we could perform a similar first to fifth (or more) loosening torque study and similarly a plot of loosening torque signature. The value to the reader is to know that these are available engineering techniques to provide a better understanding of an assembly's reusability.

We could combine the result of all of these torque applications, namely assembly clamp load, to access reusability. It is reasonable to expect our assembly to have some decrease in clamp load, hopefully small in magnitude, with repeated loading and loosening cycles due to localized permanent deformations of hub, insert propeller, and bolt. This "bedding-in" effect is one of the reasons for the frequently used maintenance instruction of periodic retightening of assembly fasteners. Reviewing our reusability factors, we can list tightening torque, loosening torque, service time, and clamp load. These are our important assembly factors to assess and understand reusability performance.

Next, we will turn our attention to the strength of this propeller assembly. A usual distinction can be to divide assembly strength into static and dynamic strength. We mentioned assembly clamp load in the previous paragraph. Clamping load, measured in Newtons or pounds measured in metric and inch system units, respectively, can be considered a static strength property of the assembly and is typically given the term preload. A useful measure of static strength is the ratio of the strength of the assembly as a unit compared with the material strength of the individual assembly members. Turning our attention to those, we can use the yield and ultimate tensile, compressive, and shear strength properties of the materials used for each assembly member.

As it is the prime fastening product in this assembly, let us evaluate the static strength of the 12-point bolts used. We can look at both the raw material certificates of the steel used to obtain the chemical and physical properties of the raw material, and we can similarly obtain the yield and tensile strength, elongation, and reduction in area and hardness of the bolts. Each individual bolt will have some small variation within specification of these properties. Each should be depended upon to meet the specification minimums.

A good way to think of these strength properties is as hold power potential. For a 12-mm diameter bolt, it is reasonable to expect a yield strength in the 900 MPa range and an ultimate tensile strength of 1000 MPa. For the tensile shear area of an M12 × 1 6 g screw thread, this would give us an area of approximately 101 mm^2 I obtained this estimate by finding the circular area of a diameter around the pitch diameter, which gives a useful approximation for applications work. The precise thread stress area is often calculated by averaging the pitch and minor diameters, and I have considered it as a circular area inclined along the plane of the lead angle. In this application, the load at yield would be approximately 91,437.7 N, or 914 kN. For assembly design review purposes, we would make our design calculations based on a percentage of the 900 MPa yield strength. Up to 75% of bolt yield is one preload target sometimes used, depending on the assembly. Turning our attention to the strength of the composite propeller, its compressive strength will be a testable, known factor. Metallic inserts may be in place to reinforce bolt through holes and bearing areas. The aluminum hub will have yield, tensile, elongation, reduction in area, and hardness values. It is reasonable to expect this to be lower than the steel bolt. For this reason, high-strength screw thread inserts are installed in the internal hub threads. The inserts have strength properties, as evidenced by their chemical and physical raw material properties, and possible heat treatment and hardness values, that approach or exceed those of the mating bolt. Taken together, they provide a good assembly site to clamp the propeller to the hub with the bolts. We can measure the strength of the materials and then test the assembly as a unit to develop an index of the assembly and material strength. The index will give us a good ratio of fastener efficiency. Welds, for example, have joint strengths higher than the materials being welded. Reusability is understandably low. Our propeller assembly has good reusability. An index of 60–80% of the lowest static strength material, perhaps the aluminum hub, is within reason.

Finally, we offer an opinion about this assembly's appearance factors. A good visual can be had by standing back, taking a good long look at the propeller assembly, and forming an opinion on how it looks. By what criteria do we judge appearance? Seeing numerous propeller assemblies could serve us well. As could continuing education and training in industrial design. If it were a consumer product, focus groups have proved both useful and quantifiable.

As a general observation, the composite material exudes a high-tech, advanced look. Our 12-point bolts are often found in aerospace as well as in high-performance and racing automobile applications. The aluminum can provide a bright, polished, and aerodynamic look. We could draw the subjective opinion that on appearance criteria we would score well.

Turning our attention to dynamic strength, I would like to offer a real-life application as an example. The screw was a special socket head cap screw with a head that was machined to provide a barrel-shaped outer head surface, but a full bearing area. It was an inch system screw with material and mechanical properties very similar to our example above. My involvement as an employee of the manufacturer of the screw was to follow up on the fatigue failure of a bolt from a lot manufactured by my employer.

The screw, used in an aircraft, had broken during a flight. Fortunately, the aircraft landed safely with the remaining bolts still intact. As I recall, we were provided with the fracture threaded section, which was removed from the hub and sent by the propeller manufacturer as a corrective action. We discussed static strength earlier in this section. Dynamic strength, of which fatigue strength is an important one, was at play here. The fracture plane of the remaining threaded area presented the classic fracture planes of a fatigue failure.

Beach marks are so named as they appear like the surface of beach sand lifted by waves of an incoming tide. Metallurgical examination of the material showed it to be within print specifications. A dimension examination was performed from unused screws still remaining from the lot. No discrepancies could be found, yet the fatigue failure was obvious. The cause turned out to be the installation of the propeller on an engine that was supercharged and flying in a high-altitude, rugged terrain, greatly increasing the dynamic loading. The screw head fractured from too high an alternating service load, carrying past the screw and propeller assemblies endurance limit. A redesign of the screw increased the fatigue strength as measured by its endurance limit by increasing the fillet radius in both dimensions, reconfiguring the head's outside diameter, and rolling the screw threads with unified "J" profile threads after heat treatment.

For our next application, we will review a class 5 threaded alloy steel stud threaded into an Inconel 718 valve body. The threaded stud serves to hold the valve bonnet onto the valve body. It needs to be reusable, but with the additional requirement that the stud-threaded end in the body must not be subject to loosening from the dynamic service loads of the valve opening and closing on high volumes of flowing fluid in the valve body. From a strength review of the stud, valve body, and valve bonnet assembly, the studs must be capable, when installed and tightened using high-strength nuts over engineered washers, of carrying substantial tension loads exerted by the fluid pressure in the valve.

Each stud's material and geometry will retain the chemical and physical properties after heat treatment to meet its specification requirements. As a circular stud pattern installed around the valve body flange, they will, in combination, provide a design factor multiple of the maximum service loads anticipated for this assembly.

While the class 5 interference screw threads on one end of these valve body studs are relatively sophisticated, their appearance factors are about as straight forward as its gets. The studs are unseen except for the several turns, or thread and point length of the standard unified (UN) screw thread protruding past the top face of the nuts when installed or torqued to design preload.

With respect to reusability, the stud and nut machine screw threads should provide large numbers of tightening and loosening cycles with none but the smallest of local deformations, almost not measurable with advanced metrology equipment. The class 5 screw threads in the valve body will require high, significant levels of loosening torque to both install and remove. This high on-and-off torque is the result of the class 5 interference screw threads. By design, their pitch diameter is larger, by a small amount, around 0.001″, than the mating internal threads. Our measure of strength in this application includes, in addition to the typical yield and tensile strengths, the resistance to vibrational loosening of these studs in service. Care is required in the dimensional control of both the class 5 fit stud threads and the mating internal screw threads. This application developed because it required trouble shooting as a corrective action to what had become an expensive problem. The class 5 alloy steel studs would thread part way into the tapped holes and then bind up. They could neither be driven further into engagement nor removed. The class 5 screw thread fit relies on the control interference of the thread flanks with a datum at the pitch diameter cylinder. The major diameter is relieved, or reduced to provide interference relief at the engaged screw threads major diameters. In this application, a judicious measurement using system 22 pitch diameter and functional diameter measurements were made of both the studs and more importantly the internal screw threads tapped into the valve body. The internal screw threads were too undersized and too small to provide the intended class 5 fit. A tap of the same diameter and threads per inch but with a slightly bigger "H" number was used to tap the holes. This now correct internal thread size accepted the proper class 5 alloy steel studs with the signature of driving and setting torques expected. As a point of interest, the class 5 studs were the initial suspect of the expensive assembly problem. This is very common when fastener problems occur. It is very easy to jump to conclusions and blame one fastener or another. The lesson that can be taken from this application is to view the assembly as a system. In seeking to understand its fastening performance, it is a good advice to consider each component as part of the whole system's performance; inspections and tests can then be thought through and run to test each carefully understood validity. The proof of a good solution, and a good fastening design will be the repeatability of the expected fastening performance—this leads to robust assemblies, and it is hoped, satisfied customers. Before we move to the next application, a review of the mentioned "H" number of the screw thread tap should be instructive.

In screw threads, a useful metrology term is basic size. Basic size is defined as the maximum material condition of the internal thread. In this application, knowing the basic size, thought of as a circle with external thread inside the circle and the internal thread around the circle, allows us to visualize the interference. Screw thread taps, in addition to being starting, tapered, or bottoming in terms of the thread length on the tap shank, are sized by diameter in terms of the base circle for their diameter and pitch. Zero equals the base circle, with each whole number representing half a tenth (inch system) or 0.0005". The tap is graded as G for ground, as are most taps as opposed to being C for cut. H1 would be 0.0005" over basic, GL1 would be 0.0005" under basic, GH2 would be 2 × 0.0005" = 0.001" over basic, as measured at the pitch diameter. Similarly, a GH5 tap would be designed to cut a screw thread that is 0.0025" over basic size.

Our next application is at the larger diameter range of the size universe. A very large truss bridge used a common pin and hanger support system for the roadway. A good visual of a pin and hanger assembly is the rollers and side plates of a chain such as found on a bicycle, motor cycle, or chain saw. The side plates carry the loads applied to the roller pins, which have a specific diameter, width, and pitch between rollers. This geometry of the side plates along with material properties for the pin and hanger strength are critical to joint integrity. A working bridge exposed to environmental elements can degrade in strength. This is especially true in the absence of proper maintenance. Microcracking in the pin and hanger plates have caused failure of these bridge roadway support systems. For this reason, supplemental support systems have been retrofitted on bridges. In this application, stainless steel rods are threaded on each end. One end of each of four rods is threaded into threaded couplings, to make a very long column with threaded ends. The ends of the assembled columns are used to hold spreader beams above and below the main tower of the bridge. Each bridge tower has four of these stainless beam columns. Each pair of columns clamp the upper and lower spreader beams. This assembly of rods, couplings, and spreader beams form the backup support system, which was retrofitted onto the bridge in the event of a pin-and-hanger performance problem. The tightening of the rod columns is accomplished using flange and jam nuts. The tightening in the flange nuts was performed hydraulically. The tension was applied to the rods to elongate them. Once at the calculated preload, the right-hand threaded flange nuts were rotated to seat them. Then, the hydraulic pressure was released, applying the clamp load to the spreader beams, bridge towers, and roadway. Preload was 500,000 pounds. The threaded joints were required to pass a preproduction tensile test. The proof load for the test requirement was 4,000,000 pounds. The threaded rods and nuts were required to withstand an applied load of 4,000,000 pounds without failure. During the initial preproduction proof load test, the internal screw threads stripped at a load of approximately 3,800,000 pounds, failing to pass the proof load requirement. Three organizational groups were in dispute with respect to the root cause of this failure. The manufacturers of the thread rods, nuts, and couplings made the claim that the design was not capable of meeting the 4,000,000 pound proof load requirement. The rods were being manufactured by one company with the equipment to handle 30 foot long, 7.5" diameter rods, while the internally threaded nuts and coupling were being machined from round bars at another manufacturer. It happened that they are both located in the same town, a few miles apart. The civil engineering firm, having responsibility for the design countered that the design was capable and that manufacturing error was the cause of the pre-production failure. In the center of the dispute was the bridge authority with a retrofit construction project of many millions of dollars stalled in preproduction testing. My employer was located by the civil engineering firm by an internet search of screw thread technology. They posed the questions, should the designed rod and nut have successfully completed the proof load test? Was the design capable? I was given the project with the goal of answering this question.

Three fundamental equations answer a lot about threaded fastener performance. In this application, the equation $S = P/A$, stress equals load divided by area, was the key equation needed to answer the design capability question. To be completely impartial and technically correct, the areas of both the internal and external screw threads were calculated for the loosest fit that would be possible and still be within specification. The external rod threads by design were required to be class 2A and were calculated at the smallest dimensions possible and still are within class 2A tolerances. Similarly, the internal nut screw threads were calculated at the largest possible dimension possible while still being within class 2B tolerances. This combination provided a threaded joint at minimum material condition for both rod and nut screw threads. This gave us the basis for calculating the A, for area in our equation. Next, we found the S, for stress. To determine proof strength, the yield strength of the nut and rod materials were used. The nut material's yield strength was 55,000 psi. This was lower than the rod's by a significant amount. The differential in yield strengths was in the magnitude of 30,000 psi or almost 55%. As a rule of thumb, a strength differential of 15%, with the internal screw thread material being lower combined with a length of engagement of 0.85 times the thread nominal diameter is sufficient to have the external component yield before the internal thread fails from stripping, provided the thread geometry is within specification dimensional tolerances. The factor that is in the external thread stress area equation is the external thread pitch diameter. The factors that are in the internal thread's shear area are the internal thread pitch diameter, the mating external thread's major diameter and the length of thread engagement. The overlapping pitch and major diameters define the screw thread pressure flank area.

The internal thread's shear area times its material yield strength equaled the load in pounds at yield, $SA = P$. This calculation indicated that the threaded joint should have met the 4,000,000 pound proof load by a margin of 400,000+ pounds. The design was capable of meeting the proof load test requirement.

A subsequent set of test parts were manufactured using system 22, indicating screw thread gages to verify specification compliance. These subsequent parts passed the proof load test with room to spare. Subsequent antidotal discussion indicated that to facilitate assembly of the previously failed parts, hand filing of the test parts screw threads was performed to clear screw thread interferences which prohibited the initial assembly of these parts.

This validated the importance of maintenance of the correct screw thread geometry on pressure flank contact area.

If the last fastening application was large, the following is at the opposite end of the size range. When brain surgery is performed, part of the skull is sometimes removed. When this skull cap is replaced, as can be visualized, it must be placed with surgical precision. One of the fasteners used in the surgical replacement is a 0.060-80 UNC-3A threaded component, with a special shaped drive and an effective thread length of 1.875″. This 80 threads per inch screw is mated with an internal thread component, which form part of the device placed on and above the patient's head. The point of each of these screws enables the surgeon to apply force through the axis of the screw and position the skull cap back into position. Dividing 80 threads per inch into 1.0 in. we find a lead of 0.0125″ for each 360 rotation of the screw. This provides micrometer-type adjustment to the pressure. In this application, cumulative lead error was of critical concern, especially in this longer screw. From one pitch of the thread to the adjacent pitch, lead variation may only amount to linear variance in the fifth decimal place, or tens of millionths of an inch. However, for 1.875″ this adds up and can cause binding in the clearance of the screw threads at class 3A tolerance and the internal thread at class 3B tolerance. Increasing this clearance, perhaps by increasing the screw thread radial clearance to 2A and 2B, respectively, was not an option due to the increased radial clearance decreasing the axial positioning of the screw point on the skull at the point of application. Screw thread measurement techniques utilizing the differential of pitch and function diameter screw thread measurement as an indication of lead variance were introduced successfully in the manufacturing of these screws. This enhanced screw thread inspection equipment enabled 0.060–80 UNC-3A screw thread lead variance control to a lower magnitude over the length of engagement of the mating part. Very long thread lengths relative to nominal screw thread diameter, such as in this application, can require specific attention to the effects of cumulative lead on screw thread function. It can be noted that micrometers, with a requirement for tight control of lead, are typically head to adjustment lengths of 1.00″ length. For example, micrometers come on zero to inch lengths, three to four inch lengths, but not on zero to four inch lengths. Control of cumulative micrometer lead screw variation over lengths of 1.00 in. for these small diameter screws is one of the factors.

As mentioned in the preface to this second edition, each chapter will introduce an engineering topic of importance to enable the reader to develop a better understanding of the fundamentals of fastening. At its root, mechanical fastening is both an art and a science; a person needs both to advance good engineering practice. For some, the study of mechanical engineering may have been confined to classroom experience and memory. Some may need a reminder or refresher on principles, problems, and solutions in the past studied and, in the present less frequently required nor used. For others, these founding subjects to the science of mechanical engineering and the fastening arts will be perhaps new material. For both of these polar opposites, and the continuum of technical understanding lying within, each of the nine chapters in this second edition of *Mechanical Fastening* will present a topic that will help with the science of understanding, presenting, and solving joining and assembly problems. We will start with fastener statics.

1.5 Fastener Statics

As its name infers, statics deals with forces and reactions at rest. This rest is properly named equilibrium. Equilibrium is defined as a state of balance or equality between opposing forces. I have found this understanding of forces in balance to be at the core of being a good fastener application engineer. As proof, I have observed that fastener failures and assemblies providing less than satisfactory assembly performance are precipitated by forces out of balance. Consider the previously referenced propeller hub bolt and special socket cap screw application. By turbocharging the internal combustion engine, the application force was increased. The strength of the opposing and resisting socket cap screw force was not. These forces, which were in balance and equal on previous aircraft engines, lost the equality. The imbalance caused a fastener failure.

In the fundamentals of fastening, forces can be thought of as tension, compression, and shear. The forces are vectors having magnitude and directions. In SI units, the magnitudes are measured in Newtons and kilonewtons. In the inch system, pounds and tons of force are used. Vector directions in each system are measured in direction. These concepts of force vectors and fastener forces in equilibrium are important for a more complete understanding of fastener performance.

In the previous section, the bridge support application was presented composed of large-diameter stainless steel rods, couplings, spreader beams, and nuts. Working in that project, I studied as many texts and technical articles that I could locate on the subject of bridge engineering. One of the things that made an impression on me was the civil engineer's ability to "read" a bridge. By that, it was meant that an engineer with a good technical understanding of bridge engineering could observe a bridge and read the direction and magnitude of the forces and reactions of a given bridge. It seemed to me that anyone wanting to become proficient in fastener engineering should similarly be capable of reading an assembly and discerning the forces, reactions, magnitudes, and directions of any given fastened assembly.

In fastener statics, the key point presented will be the concept of equilibrium, vector mathematics, friction, bending, turning moments, and center of gravity. For each of these statics topics, we will offer specific fastener-related applications.

The deck of playing cards in Figure 1.1 at the beginning of this chapter shows forces along the *x*-, *y*-, and *z*-coordinates. Forces that are drawn are being applied up, down, and sideways. The text presents the simple idea of fastening them with a rubber band. Let use that to define static equilibrium. Of side forces in these three, tension, compression, or shear, directions can work to move the cards out of assembly either singly or in combination. The rubber band can oppose any of these forces if it has the capability. If the applied forces are opposed by equal and opposite reactions, no motion occurs. The cards are secure and the assembly is said to be in static equilibrium.

The concept of static equilibrium is the key to understanding how fasteners and fastened assemblies perform. I find this concept of static equilibrium to be a key concept. Most people view an engineered assembly in everyday life and take for granted the forces and reaction present. We use our plumbing fixtures, drive our cars or use public transportation, ride elevators, and pick up mobile devices and can be oblivious to this balancing of forces that keeps things working, at static equilibrium.

Let us consider the static equilibrium of an assembly many orders of magnitude larger. A jet airplane sitting at the boarding ramp at an airport is supported by its landing gear, at equilibrium. On the wings of that aircraft are mounted jet engines of considerable weight. The force of gravity pulling these engines downward is proportional to the mass of the engine, the mass of the earth, and the distance between their centers of gravity. Indeed, these are the factors that mathematically define the force of gravity. Since we observe that the engines are in fact well attached and stationary as the plane rests before its next flight, reactions by the assembly are equal and opposite to the force of gravity as well as any other forces such as wind, fueling, or other outside forces that may be applied. It is hoped that each of these nine chapters will not only provide knowledge of mechanical fastening, but also provide a good grounding or refresher on the fundamental engineering which provides its technical support. When you learn to read a bridge, or any mechanical assembly, evaluate the forces and reactions which compose or disturb its static equilibrium.

Turning our statics attention to vectors and vector mathematics. Whole engineering texts are available on the vector algebra and more advanced numerical techniques. For our statics overview, we have said that vectors have both magnitude and directions. What are the units of measure? In SI units, it is Newtons and kilonewtons. We will use Newtons (N) and for larger forces kilonewtons (kN). Force vectors can be added, subtracted, and resolved into rectangular components.

Let us consider a system at equilibrium, with a fastener clamp load of 4500 N, or 4.5 kN, being opposed by a 4.5 kN reaction from the assembly. If we add a +4.5 kN and a −4.5 kN to resolve our forces, we get zero, or equilibrium. If we pull on the assembly with an additional 1 kN, the reaction must increase, if the assembly is capable, or we will no longer have static equilibrium, and motion will occur. Force vectors in line are called collinear. In this example, we could say that the magnitudes were, 4.5 and −4.5 initially, with directions of 0 and 180°.

If we have a force at 45°, let us say a 5 kN force at 45°, we could resolve this force into rectangular components. We would multiply the cosine of 45°, 0.7071, to obtain the horizontal component of the force, 3.5355 kN and multiply the force by the sine of 45°, 0.7071, to obtain the vertical component. In this way, forces in different directions can have their x, y, and z components resolved and then added or subtracted to obtain the resultant. This can also be drawn graphically by carrying horizontal and vertical lines from the forces to the x- and y-axis.

Figure 1.13 shows a primary tensile force at zero degrees, or vertical, and a smaller amount of force components as they are transferred by the socket head cap screw through the head bearing area and the pressure flanks of the screw threads. As most fastener threads are of a 60° inclined angle, the cosine of 30°, 0.866, and the sine of 30°, 0.500, are useful to know in resolving pressure flank forces into radial (or hoop forces) and axial force components. If we apply significant pressure flank force on a thin-walled internal part, the radial forces are important to know.

We spoke of the parked aircraft's jet engines while discussing static equilibrium. Let us return to it to review center of gravity. The center of gravity is the point in a component, through which the force of gravity passes. A modern commercial jet engine assembly weighs between 14 and 20 kN by an educated estimate. If we take the mean, 17 kN, this force of gravity can be

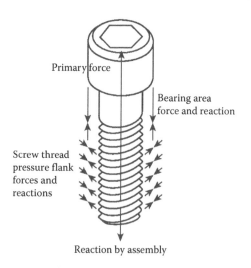

FIGURE 1.13
Socket head cap screw forces.

thought of as passing through the center of gravity of the engine assembly. The center of gravity is a useful read of an assembly. For a symmetric part of uniform material density, such as a ball, or a cube, we can trace diagonal lines from the corners or side midpoints and diameter lines through the spheres of these shapes, and know their internal point of line intersection to determine the location of the center of gravity.

Since we intuitively know that not all forces, including gravity, will always be applied this conveniently, statics provides the concept of turning, or bending moments, and a related applied force, the couple. A turning moment is defined as a force which causes or tries to cause if resisted by an opposing moment to equilibrium, rotation. Turning moment has both a force and a direction of rotation. In mechanical fastening, the torque applies to tighten a thread fastener is a turning moment.

Turning moment is a force applied at some defined distance from the axis of rotation of the part to which the turning moment is applied. While we could use any torqued bolt, screw, or nut, we will use the socket screw previously used. A hex key inserted into the screw's hex socket provides a wrench arm of some defined length. We will use 2.5 cm as our hex key's long-arm length. Turning in the clockwise direction for a right-handed screw thread will seat the screw and pull on the hex key with a force of 90 N (2.5 cm is 2.5 m). The turning moment is 0.250 m × 90 N or 22.5 N m. Torque or turning moment is expressed as a force applied through a distance. A sign of a working knowledge of engineering understanding is the proper use of terms.

When someone uses an expression of 20 pounds of torque, it shows a lack of this knowledge. A couple is a type of moment where the applied force is in the form of a pair of turning moments.

When you turn an automobile steering wheel with two hands, you are applying a couple. An example of a fastener couple would be a twin lead thread forming screw. The torque applied to drive the screw is a moment. The turning force at the point of the screw is a couple.

Friction is a statics concept, which is fundamental in the mechanics of screw thread function. Referring to the thread pressure flank vectors, these forces are the result of the clamping forces, developed axially in the shank of these fasteners, generated by the torque applied to the wrenching surface. All of these metal surfaces sliding against each other result in friction. Statics provides a mathematical tool to calculate the amount of friction in a fastening system. The statics term which is used is the Greek letter mu. The formula is applied force equals mu times the normal force. To demonstrate friction using statics, let us place a 5 kg steel block on a dry steel plate. The force holding the block on the plate is the force of gravity, 5 kg, through the center of gravity. This is the normal force.

The force required to move this right or left is mu times the normal force. Mu can be found experimentally by testing on any application and also can be found as an estimate from standard published values of mu. At the high friction end of the range, mu, which is also referred to as the coefficient of friction, can approach unity, or a value of 1.0. At these high levels of friction, much of the input Force (F) to overcome load (N) and start movement is needed to overcome friction. For example, with a coefficient of friction of 0.8, a force of 4 kg is required to overcome friction. Reducing the coefficient of friction to 0.4, or cutting it into half, reduces the force to start motion to 2.0 kg. Using the vector mechanics we introduced above, if we incline the surface 4–5°, we can use rectangular coordinates and the sine, cosine, and tangent functions to find the frictional forces at work on a block under gravity load on this inclined plan. This is useful for our work with fasteners using screw threads. Screw threads can be thought of as inclined planes wrapped around cylinders. If we unwrap a single lead screw thread, we will see an inclined plane. The angle of inclination is the lead angle, and the block is the thread flank clamping load. We can determine the lead angle by reference to thread specification tables and numerous handbooks. It can also be calculated by taking the arc tangent of the lead divided by 3.1416 times the pitch diameter.

The concept of static friction is very useful in thread fastener applications. The input torque is converted into clamping load through the diameters of the fastener and requires the overcoming of friction to develop the load. The higher the various fastener surfaces coefficients of friction, the more torque that will be required to develop any given clamp load. In Figure 3.7 shows a graph plotting the torques to achieve a target clamp load in a specific threaded fastener with five different frictional coefficients.

The equations used, sum up mu for the fastener materials into an index, K, which can be thought of as the index of friction. In my practical experience, K values as low as 0.05 for super-lubricated parts up to K values approaching 0.35 are typically encountered in industrial fastening. It is conceivable that much high K values could be experienced in applications where the screw threads are exposed to service environments precipitating corrosion between mating fastener screw threads.

We will close our presentation of fastener statics with the concept of degrees of freedom. In engineering statics, components can be either fixed in location, such as a bolt, washers, and nuts clamped tightly through two ridged parts, or can be free to move. An example of freedom of movement would be the screw threads of the same bolt and nuts before tightening. Can the screw threads be moved up and down relative to each other? If so, that is two degrees of freedom. Are they side to side? Canted right to left? The degree of freedom is an important engineering consideration. Freedom of movement is a concept in the engineering section concluding this second edition of Chapter 2, fastener dynamics. As a final pointer on the use of statics in fastener applications engineering, the use of simply drawn free body diagrams should be a tool in the kit of anyone seeking a better understanding of the fastener mechanics. In conjunction with the ability to "read" an assembly, the line drawing of vector with magnitude and direction of the forces and reactions in any given assembly of components is a skill that will serve the reader well. An example can be found in Figure 3.5b if the reader would care to skip ahead before returning to Chapter 2.

Reference

1. Speck, James A., P.E. 1995. Thread pitch's influence on assembly speed, *American Fastener Journal*, Powell, OH, p. 52.

2

Fastener Types and Their Production

If you are responsible for the selection of the fasteners of your products and assemblies, you face an industrial fastener market with a confounding variety of designs. Similarly, if you are responsible for developing the joining techniques and procedures, you are approaching a task that can have a profound effect on the market success of your company's products versus its competitor's offering as well as legal ramifications should your work be less than effective. And finally, to undertake the layout of an assembly process, be it a new design or the modification of an existing assembly process, the decision made can have a lasting impact on the growth or decline of your process. The outcome is dependent on the efficiency of the assembly process and fastening hardware.

One way to think of these decisions is to separate them into hardware and software. The software consists of the overall assembly and joining engineering approach, complete with assembly drawings, assembly procedures, and instructions, as well as inspection and testing documentation. The hardware would be the actual fasteners, installation tooling, and all of the equipment required at the assembly site. By separating the assembly layout task into hardware and software, we can approach the implementation of hardware selection based on the approach predicted by the software. To carry this analogy just one step further, we can make more informed decisions regarding the optimal assembly approach for our assemblies if we have a full knowledge of the hardware available in the fastener industry to help us accomplish our task.

This chapter will categorize some of the more commonly used fasteners by type. It will also illustrate many of the most popular designs used in each category and provide dimensional and mechanical performance data for several designs. It is hoped that it will serve as a useful reference when decisions must be made regarding the design and style to be specified for an assembly application. Throughout this chapter, manufacturing process sections will provide a technical overview of some of the manufacturing processes used to produce the illustrated designs. With a technical grounding in these fastener manufacturing processes, you will have the knowledge and be in the position to specify both assembly processes and their specific hardware implementations to meet your joint making and assembly production needs with efficiency.

The general categories are as follows:

- Tension
- Compression

- Shear
- Adhesives

Both inch system and metric system fastener designs and data will be included along with references to consult for additional information.

2.1 Tension Fasteners

Tension fasteners are the most frequently used fastening design in the industrial world. As a design strategy, directing the service loads concentrically toward a rigid, minimum member joint clamped by a well-preloaded tension fastener can often be an optimum assembly solution. The following tension-fastener geometries show some of the range of choices in the tension-fastener category. Some of the figures are made on individual pages so that a copy can be made for your own assembly dimensioning work. The drawings were made with FastPrint fastener software.

2.1.1 Pan Head

The pan head shape is a highly used tension-type fastener which is available with a wide range of drive types. It fits well within the production range of the cold-heading machinery and tooling, which form the pan head shape at high speed. It has a reasonable, although not overly large, bearing area under the head. It is designed so that, barring any overly deep or sharp drive formed into it, and also provided that a fillet radius of approximately 10% of shank diameter is formed, and it will normally fail in the threaded area if taken to its ultimate tensile strength. This represents a generally desirable design characteristic in a tension fastener in that it is not prone to failure in the head. The pan head shape also has a good appearance. Some dimensions are given in the following tables.

Pan Head Inch System Dimensions

	#2	#4	#6	#8	#10	1/4	5/16	3/8
Head diameter	0.167	0.219	0.270	0.322	0.373	0.492	0.615	0.740
Max. to Min.	0.155	0.205	0.256	0.306	0.357	0.473	0.594	0.716
Head height	0.053	0.068	0.082	0.096	0.110	0.144	0.178	0.212
Slotted	0.045	0.058	0.072	0.085	0.099	0.130	0.162	0.195
Head height	0.062	0.080	0.097	0.115	0.133	0.175	0.218	0.261
Recessed	0.053	0.070	0.087	0.105	0.122	0.162	0.203	0.244

Note: ANSI standards for odd-numbered diameters as well as larger and miniature sizes are available in ANSI/ASME B 18.6.3.

	M1.6	M2	M2.5	M3	M3.5	M4	M5	M6
Head diameter	3.2	4	5	5.6	7	8	9.5	12
	2.9	3.7	4.7	5.3	6.64	7.64	9.14	11.57
Head height	1.0	1.3	1.5	1.8	2.1	2.4	3.0	3.6
Slotted	0.85	1.1	1.3	1.6	1.9	2.2	2.8	3.3
Head height	1.3	1.6	2.1	2.4	2.6	3.1	3.7	4.6
Recessed	1.16	1.46	1.96	2.26	2.46	2.92	3.5	4.3

Metric System Dimensions

Note: The International Standards for metric pan head screws are ISO 1580 for slotted pan head screws and ISO 7045 for cross-recessed head screws. Figure 2.1 shows a pan head blank.

Some of the more highly stressed areas of a pan head tension geometry are as follows:

- The head-to-shank fillet
- The cross-recess drive penetration or slot width and depth
- The web between the drive/slot bottom and the fillet

2.1.2 Truss Head

The truss head applies the tension developed in the fastener over a larger bearing area. It has a lower head than the pan head style, which can result in less strength in the head for certain drive geometries. It can be an application benefit if clearance for the head is limited. It also falls well within the normal cold-heading range, although since it is not as frequently used in industry, it may not always be as readily found in all sizes and materials.

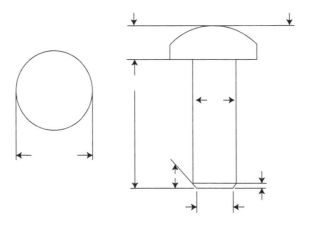

FIGURE 2.1
Pan head.

Inch System Dimensions

	#2	#4	#6	#8	#10	1/4	5/6
Head height	0.194	0.257	0.321	0.384	0.448	0.573	0.698
	0.180	0.241	0.303	0.364	0.425	0.546	0.666
Head height	0.053	0.069	0.086	0.102	0.118	0.150	0.183
Slotted	0.044	0.059	0.074	0.088	0.103	0.133	0.162
Head height Recessed	The same head height dimensions are used						

Note: Truss head punches are produced, which form a flat on the top, or face of the truss head. This gives a well-finished appearance. Also, the recess penetration depth of truss heads, as measured with a recess penetration gage, are slightly lower than the more common pan head geometry.

Metric System Dimensions

	M1.6	M2	M2.5	M3	M3.5	M4	M5	M6
Head diameter	3.3	4.9	5.7	6.5	9.7	10	12	14
Head height Slotted	0.9	1.3	1.5	1.7	2.5	3	4	4.2
Head height Recessed	The same head height dimensions are used							

Notes: The preceding metric dimensions are derived from standard truss head forming punch dimensions. Certainly, metric dimensioned truss head screws can be cold-headed and roll-threaded on an engineered basis. Additionally, cross-recessed versions of the metric truss head design are technically possible. Figure 2.2 shows a truss head blank.

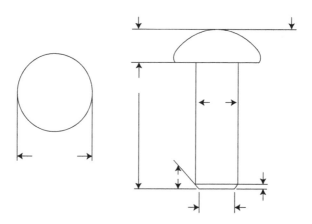

FIGURE 2.2
Truss head.

2.1.3 Hex Head

Similar to the pan head screw, hex head screws and bolts are industrial work-horses. They are used as a geometry of choice in numerous applications for their ready availability, commonality of tightening tools, and load-carrying ability. With respect to load carrying, it is important to note that hex head cap screws, both inch and metric systems, are produced and heat-treated to several strength levels, using specific raw materials and heat-treating processes. A general listing would be as follows:

Inch System	Tensile Strength	Metric System
SAE Grade 2	Low	Class 4.8
Grade 5	Medium	8.8
Grade 8	High	10.8
180 KSI	Very High	12.9

The inch grades are noted by lines on the top of the inch hex heads. Add two to the lines to determine the grade. Metric grades are often headed or rolled onto the blank.

Inch System Dimensions					
Hex Head Screws	**#2**	**#4**	**#6**	**#8**	**#10**
Across flats, max.	0.125	0.188	0.250	0.250	0.312
Across flats, min.	0.120	0.181	0.244	0.244	0.305
	0.134	0.202	0.272	0.272	0.340
Across height, max.	0.050	0.060	0.093	0.110	0.120
Across height, min.	0.040	0.049	0.080	0.096	0.105
Hex head bolts unthreaded					
Body max.	1/4	1/2	3/4	1	1 1/4
Across flats, basic	7/16	3/4	1 1/8	1 1/2	1 7/8
Head height, basic	11/64	11/32	1/2	43/64	27/32

Note: The ANSI standard for hex head screws is B18.6.3. It covers diameters from #1 to 3/8, which can either be threaded "to the head" or have an unthreaded body depending on the fastener length. Good corner definition or fill is a factor for satisfactory wrench tightening. Accordingly, the hex head across corner dimension is inspected with a gaging ring at a specified minimum diameter beyond which the inspected fastener must protrude a specified minimum distance.

The ANSI standard for hex bolts is B18.2.1 and covers diameters from 1/4 in. through 4 in. A basic thread length is specified for each diameter, beyond which an unthreaded body is typical for standard bolts. Straightness of the entire shank section, both threaded and unthreaded, can be an assembly factor, particularly on very long bolts. The ANSI B18.2.1 standard specifies straightness limits. ANSI B18.2.1 also specifies "finished" hex fasteners, which have a defined washer face.

Metric System Dimensions

Hex Head Screws	M1.6	M2	M3	M4	M5
Across flats, max.	3.2	4	5.5	7	8
Across flats, min.	3.02	3.82	5.32	6.78	7.78
Across corners, min.	3.41	4.32	6.01	7.66	8.79
Head height, max.	1.225	1.525	2.125	2.925	3.65
Head height, min.	0.975	1.275	1.875	2.675	3.35
Hex head bolts	M8	M10	M16	M20	M42
Unthreaded body diameter, max.	8.58	10.58	16.7	20.84	43
Across flats, nominal	13	16	24	30	65
Across corners, min.	14.2	17.59	26.17	32.95	71.3
Head height, nominal	5.3	6.4	10	12.5	26

Note: The ISO standard for hex head screws is ISO 4017 for product grades A and B. ISO 4018 covers product grade C. Diameters dimensioned start at M1.6 and stop at M64. Grade A is for diameters up to M24. Grade B is for the larger diameters. Product grade C covers hex head bolts, as does ISO 4014, which also covers from M1.6 to M64. A related standard, ISO 4015, specifies hex head fasteners with a reduced shank diameter. The reduced shank diameter gives more total elongation for a given preload and is a desirable assembly feature in some applications. Additional ISO standards such as ISO 7411 and ISO 7412 cover high-strength structural bolts with larger width across flats and correspondingly increased under head-bearing areas.

Since the metric thread system uses a preferred thread pitch, which is sized partway between the unified inch system coarse and fine threads, ISO 8676 supplies dimensions for fine pitch metric hex heads. Figure 2.3 shows a hex head blank.

2.1.4 Socket Screws

Socket screws are high strength, primarily smaller-diameter head tension fasteners produced with an internal wrenching tightening drive. They are produced to industrial standards that are in some areas more tightly toleranced and controlled than other tension-type fasteners. They were originally

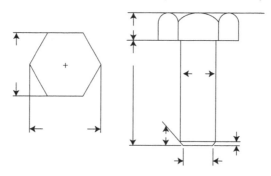

FIGURE 2.3
Hex head.

designed for tool and die work applications but over the years have been found in many applications where their high strength and compact size offer application benefits.

The American national standard is ANSI B18.3 and the International standard is ISO 4762. Variations of the socket screw design use spine or six-lobed internal drives. In the international ISO, and its predominant German DIN socket screw standards, several steel socket screw strength levels are available, grade 12.9 being the prime one. Others with a lower strength can be supplied, so it is good practice to check the head markings for the strength grade. In the inch standard, a nominal 180,000 psi tensile strength, depending on the screw diameter, is standard with through-hardened alloy steel being the fastener material used. In normal practice, a black oxide or thermal finish is applied, which improves the esthetics but provides little corrosion resistance. In tool and die work, this normally presents no problem. In more environmentally demanding applications, this can be a factor.

In both inch and metric series, socket screws can be supplied, which are made from austenitic 300 series stainless steel. While this does improve corrosion resistance, tensile and yield strengths are significantly lower. An additional application design factor is drive access and durability. While the compact, internal drive feature of socket fasteners is an application benefit, buildup of debris as well as the use of poor quality or ill-fitting hex wrenches can limit torquing ability.

Inch System

	#4	#6	#8	#10	1/4	1/2	3/4
Head diameter	0.183	0.226	0.270	0.312	0.375	0.750	1.125
	0.176	0.218	0.262	0.303	0.365	0.735	1.107
Head height	0.112	0.138	0.164	0.190	0.250	0.500	0.750
	0.108	0.134	0.159	0.185	0.244	0.492	0.740
Socket size	3/32	7/64	9/64	5/32	3/16	3/8	5/8

Metric System

	M2	M3	M4	M5	M6	M8	M10
Head diameter	3.98	5.68	7.22	8.72	10.22	13.27	16.27
	3.62	5.32	6.78	8.28	9.78	12.73	15.73
Head height	2.0	3.0	4.0	5.0	6.0	8.0	10.0
	1.86	2.86	3.82	4.82	5.70	7.64	9.64
Socket size	1.5	2.5	3	4	5	6	8

Note: Heads can be supplied knurled or plain. The metric sizes usually have the strength grade marked on the head. Some manufacturers use an identifying knurl as a manufacturer's identification. Figure 2.4 shows a socket head cap screw blank. A similar design with an alternative drive would be an internal socket with six lobes. For tamper resistance, either socket drive could have a pin formed from the base of the socket, which would require a matching wrench. An external 12-point drive with flanged, integral washer feature is also used for high-strength applications.

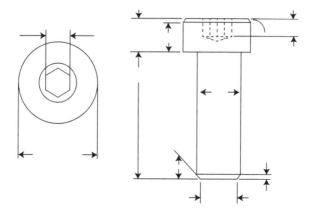

FIGURE 2.4
Socket screw.

2.1.5 Blind Rivets

Blind rivets are also tension fasteners although the loads they are designed to carry are of a much lower magnitude than socket products. One of their prime benefits is speed of installation as they do not require threads and torquing to make up their assemblies. They are made of two assembled parts, an outer rivet body, which remains clamped in the assembly, and an inner mandrel, which is drawn up tight with a riveting tool, either hand powered, or more commonly power actuated. The mandrel body separates by design during the installation by snapping off. Blind rivet bodies and mandrels can be manufactured and supplied from several different material combinations, with steel and aluminum being the most common.

2.1.6 Some Dimensions for Blind Rivets

	Inch System				
	No. 3	**4**	**5**	**6**	**8**
Body diameter	0.096	0.128	0.159	0.191	0.255
	0.090	0.122	0.153	0.183	0.246
Mandrel diameter	0.057	0.076	0.095	0.114	0.151
Hole size	0.100	0.133	0.164	0.196	0.261
	0.097	0.129	0.160	0.192	0.257

Note: Blind rivets come in a variety of grip range and rivet/mandrel material combination and head styles. Figure 2.5 shows a blind rivet with mandrel. Blind rivet tools are available in a wide range of designs from simple hand-actuated tools through power-actuated guns, which can be counterbalanced above an assembly site to automatic feed systems that can operate in a high-production environment. It is good practice when considering blind rivet tooling to factor in tool maintenance as well as periodic cleaning and replacement of wear parts of the tooling such as mandrel grippers.

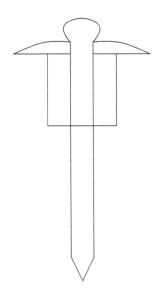

FIGURE 2.5
Blind rivet.

Some of the significant areas of blind rivets are as follows:

- The mandrel notched dimension
- The rivet body wall thickness
- The rivet bearing area

2.1.7 Self-Clinching Studs

Self-clinching studs and related self-clinching hardware use lugs formed under the head to press into sheet metal components to provide a tension-fastener attachment. They also have a radial groove between the lugs and the machine screw threads for the sheet metal to snap into. The lugs prevent rotation of the stud, while the groove resists push-out forces.

	Inch System Dimensions	
	Installation Hole Diameter	Min. Sheet Thickness
#2	0.085	0.040
#4	0.111	0.040
#6	0.137	0.040
#8	0.163	0.040
#10	0.189	0.040
1/4	0.249	0.062
5/16	0.311	0.093

Metric System Dimensions (mm)		
M2.5	2.5	1.0
M3	3.0	1.0
M3.5	3.5	1.0
M4	4.0	1.0
M5	5.0	1.0
M6	6.0	1.6
M8	8.0	2.4

A typical clinch stud installation will use an anvil on the sheet metal site opposite the side the stud is being installed from in order to provide a reaction to the stud as it is pressed into the sheet. The anvil can be constructed from a solid tool steel section with a through hole to just clear the thread major diameter. Self-clinching hardware is available in a wide range of designs and materials. Both externally and internally threaded fasteners are offered, which can be installed in sheet applications. They can provide an efficient alternative to "loose" hardware by making the fasteners "captive" to the site. In several applications familiar to the author, self-clinching studs are automatically pressed into large vending machine frames and the resulting structure is powder coat-painted for corrosion resistance and appearance prior to assembly, with the threads coated with a plastic that prevents paint adhesion but still facilitates the final component assembly. Figure 2.6 shows a self-clinching stud with the underhead lugs, which provide an antirotation feature once installed; the annular groove is for pushout resistance.

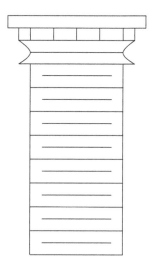

FIGURE 2.6
Clinch stud.

2.1.8 Weld Studs

Weld studs are fastened to a sheet of metal using welding current to fuse the fastener metal to the site's metal. These provide a solid attachment mounting for subassemblies and components. They are generally cold headed and roll threaded. They can be obtained in several types of welding heads, as follows:

Type U3   Three weld projections are on the underside of the pancake-shaped head. The threaded end protrudes through the sheet metal.

Type T3   Three weld projections are on the top of the pancake-shaped head. Both the head and the threaded end are on the same side of the sheet metal.

Type TD   The pancake-shaped head has a single dome-shaped weld projection in the center top of the head. Both the head and threaded end protrude from the same side of the sheet metal.

Type UR   The pancake-shaped head has an integrally formed, annular weld ring formed on the underside of the head. The threaded end protrudes through the sheet metal after the stud is welded into the site.

Type US3   Similar to U3 with the exception that the three underhead weld projections are sausage-shaped in a curve, concentrically around the shank.

Type UC4   Four hemispheroidally shaped weld projections are formed on the underside of the head. The stud protrudes through the sheet metal after welding.

Type TC4   Similar to UC4 except that the four hemi-shaped projects are on the head top and head and threads are on the same side of the sheet metal after welding.

T3 Weld Stud Dimensions

Sizes	Head Diameter	Weld Projection	Height Head
4	0.260/0.240	0.017/0.013	0.046/0.036
6	0.323/0.303	0.022/0.018	0.052/0.042
8	0.385/0.365	0.027/0.023	0.068/0.058
10	0.448/0.428	0.032/0.028	0.068/0.058
1/4	0.575/0.550	0.042/0.038	0.083/0.073

Manufacturer's Standard Steel Gages

Gage #	Sheet Thickness (in.)
20	0.0359
19	0.0418
18	0.0478
17	0.0538
16	0.0598
15	0.0673

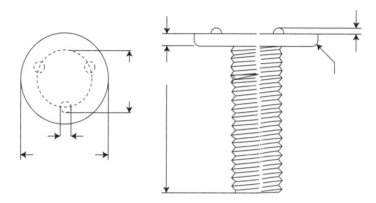

FIGURE 2.7
Weld stud.

Figure 2.7 shows a typical weld stud.

2.1.9 Grooved Pins

Grooved pins are fasteners which do not require torque to install. They are cold-headed parts with a segmented split around the shank. They can also be headless and provide shear and hinge functions in assembly. As a tension fastening, they are usually manufactured from a cold-headed blank with roll-threaded split knurls. The splits are an innovative feature in that they are separated longitudinally on the blank. A range of head styles such as round, pan, and truss head are produced in common fastener materials of low-carbon steel and cold-heading types of corrosion-resisting stainless steels such as type 302HQ, also referred to as UNS 30430 and XM-7. Diameters from a small size #0 (often referred to as an "ought") up to 1/4 in. diameter are manufactured, usually in fractional inch lengths. Figure 2.8 shows a pan head grooved pin.

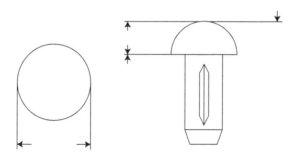

FIGURE 2.8
Pan head grooved pin.

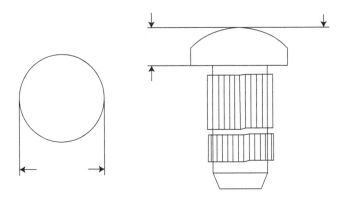

FIGURE 2.9
Split knurl drive screw.

2.1.10 Split Knurl Drive Screws

Split knurl drive screws also allow nontorque tension assembly. The straight 60° knurls are offset by half a pitch (knurl pitch is the distance from one knurl crest to the adjacent). This offset gives a degree of anchoring in the assembly hole site material. A typical split knurl design is shown in Figure 2.9.

2.1.11 Type "U" Drive Screws

Type "U" drive screws are also push-in types of tension fasteners. Type "U" threads are cold headed and roll threaded and have a helix angle of 45–65°. This high helix angle allows them to rotate as they are being pushed into the assembly site. They can be obtained in sizes as small as #00 in standard steel and stainless steels. The low-carbon-steel type "U" screws are typically carburized and electroplated. Figure 2.10 shows a round-head type "U". Flat countersunk heads with 82°, 90°, or 100° head angles could also be cold headed for a flush installation.

2.1.12 Self-Threading Screws

In applications where the reusability of a threaded fastener would be useful but the precision of separately generated internal threads is not, self-threading by the fastener as it is being installed can be a tremendous assembly cost saver. If not planned with respect to fastener performance, it can also be a tremendous generator of assembly problems.

Self-threading fasteners can be categorized into thread cutting, thread forming, and thread rolling. Considering them in order, we can evaluate their differences and similarities.

Thread cutting threads—The use of self-threading screws by cutting an internal thread is a logical continuation of the method used by most taps in

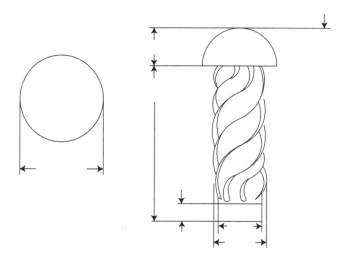

FIGURE 2.10
Type U drive screw.

generating a similar internal thread. A cutting flute or flutes on the leading point threads removes material from the wall on the assembly site hole as it is turned into the assembly. The thread helix and assembly torque overcome wall strength and displaced area. Common types are types F, 23, 25, and 17. The former has flutes rolled on during threading, whereas the others cut afterward in an operation called shank slotting.

Thread-forming threads—Some of the earliest of these were little more than either machine or spaced threads with a chamfered or gimlet point thread. Driving torques could be very high. Common types are A, AB, and C. Types A and AB were often used in early grades of thermoplastic. Type C was used in low-strength metals such as aluminum and zinc die castings.

Thread-rolling threads—These are engineered, high-performance threads designed to flow the internal wall material into the desired internal thread shape. Some popular thread-rolling geometries use three or more lobes either at the point and lead in threads or along the entire shank. This can require the fastener blank, prior to rolling, to also have a trilobular or similar shape. The cost for tooling can be higher, but the benefits in lower drive torques can be significant. Other thread shape innovations have been used for forming, for example, ramps on the flanks of threads, special blank point angles for specific after-threaded lead in thread angles of attack, special diameter/pitch combinations, and compound thread flank angles.

One of the benefits of high-performance thread-rolling screws can be a lower installation cost when assembly efficiency and ease of installation are considered. Moreover, chips and slivers can either be eliminated or reduced to negligible amounts.

Self-threading factors to consider are as follows:

1. Drive torque and stripping torque—the strip-to-drive ratio
2. The ratio of hole inside diameter to outside diameter for control of radial stresses from self-threading and tightening
3. The amount of taper, edge chamfer, and out-of-roundness of the hole, which could cause starting difficulties on the assembly floor
4. The geometry and performance of the self-threading fastener drive and its ability to transmit sufficient self-threading torque without failure of the fastener drive geometry
5. The geometric and metallurgical condition of the driver bits and tightening/installation tools for smooth starting and self-threading without excess wear or stress

2.1.13 Locking Threads

It is important to include information on fastener threads, which are by design intended to "lock" when installed in assembly. Machine screw threads are manufactured to a standard of tolerances and classes of fit which normally provide for free running. Class 3A, 2A, 1A external, and 3B, 2B, IB internal are the industry standard choices. Locking threads have a tighter fit than a 3A/3B set of threads at maximum material condition. The earliest locking threads were a simple interference fit, with the external thread being slightly larger than the engaged internal thread. ANSI standards are available for 1/4–20 and larger class 5 interference fit threads. These threads can have very high driving or prevailing torque as locking thread drive torque is called.

More efficient plastic element, adhesive, and all metal-locking threads have been developed, manufactured, and marketed, which are available in a wide range of fasteners. Just as significantly, they have more efficient prevailing, first off and fifth off torques. These drive, first removal, and fifth removal turning forces are good performance indices of fastener locking thread effectiveness. Figure 2.11 shows a locking thread design.

2.1.14 Cold-Heading Manufacturing Process

Heading is a fastener manufacturing process used to produce many commonly used fasteners. The heading process is most often performed using coiled wire at room temperature, or cold, hence the term cold heading. This is in comparison to heading, or forming, performed on raw material which is heated to either below or above the metal's lower critical temperature. Warm heading is forming at temperatures below this recrystallization temperature, the additional heat increases the formability of some more difficult-to-form fastener alloys. Above the material's critical temperature, forming is more fluid; hot heading is the only manufacturing option in some fastener designs such as extremely large diameter, long fasteners or when using exotic alloys.

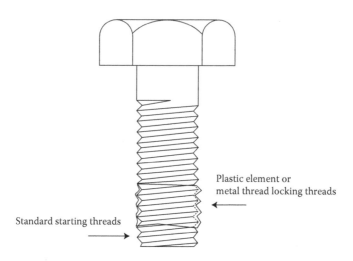

Plastic element or
metal thread locking threads

Standard starting threads

FIGURE 2.11
Locking thread fastener.

In operation, a force is applied to the end of a wire blank that has been sheared from the end of a coil of wire or bar stock, usually in the case of hot heading. The blank length is sized to provide the same cubic volume as the final fastener blank shape. The heading force that forms the material is above the material's yield strength but below the ultimate tensile strength. In this way, the wire cutoff is transformed into a fastener preform, or cone blow, while the shank is held in a piece of tooling called the die, without cracking from exceeding ultimate tensile. In a simple cold-heading machine, two punches, or blows, are delivered to the fastener blank at high speed, typical production speeds of 200–400 blanks per minute being typical manufacturing rates. The second punch again reforms the cone, or preform, into the final fastener geometry. On more advanced machinery, three or more blows are delivered to a blank, which is transferred through several dies. In heading, the die or dies form the shank, or body portion of the fastener.

The typical single die heading process would consist of the following:

1. Cutoff
2. Preform
3. Finish blow
4. Eject or kick out

Forming operations that can be performed in heading would include upsetting, or forming the head, forward extruding where a reduced shank diameter or point is formed on the blank, and back extruding where a socket or cavity is formed. Heading is efficient in its use of raw material and offers a high production speed. Figure 2.12 shows a high-speed cold header.

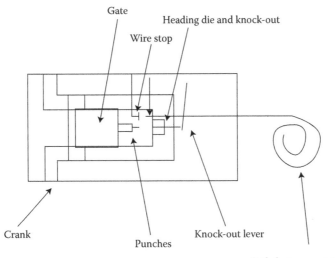

Gate

Heading die and knock-out

Wire stop

Crank

Punches

Knock-out lever

Coiled wire raw material

FIGURE 2.12
High-speed cold header.

2.1.15 Roll-Threading/Forming Process

Thread rolling, and the less well-known form rolling, are also normally cold-forming processes. They are normally used to produce the external threads on threaded fasteners. There are rotary and planetary die thread-rolling machines, but by far the most numerous are flat die roll threaders. In the flat die thread manufacturing process, a pair of hardened, tool steel dies roll an unthreaded fastener blank between their threaded surfaces. The threaded, or other form, geometry on the face of the dies are pressed into the blank as it rotates between the dies, often at speeds well in excess of 300 ppm. A practical model of roll threading would be the rolling of a pencil between the palms of your hands. By keeping one palm stationary while pressing the pencil with the other palm and moving it, the pencil "rolls." So too does the fastener blank as it is being threaded. With one shorter stationary threading die and one longer reciprocating die, a headed or machined blank inserted between the two dies at a specific location at which the threaded surfaces of both dies are matched, the blank makes a number of rotations, or turns, as it rolls through the dies. At the end of the shorter stationary die, the blank is released. It normally is allowed to fall by gravity into a discharge chute.

As can be pictured, the blank material is displaced during thread rolling. Blanks of the proper diameter are squeezed down into the root or minor diameter of the thread. The displaced material is formed up into the thread crests or major diameter. The theoretical pitch diameter is located close to what was formerly the blank or roll-thread diameter. If the threads are to be plated or otherwise coated for environmental protection after thread rolling, the roll-thread diameter and the rolling pressures may need to be planned and set up to provide a coating buildup allowance.

The thread-rolling process generates a fair amount of slippage and friction as the blank makes its work rotations through the dies. Moreover, any noncylindrical surface or concavities in the blank are transferred to the manufactured threads along with any imperfections inherent in the thread-rolling dies, their setup, and thread-rolling machine condition. Proper gaging of the rolled threads is important for satisfactory assembly of the completed fasteners.

Roll-threaded fasteners were once considered not as accurate as machined threads. With industry advances, this is no longer the case. Figure 2.13 shows a production roll threader.

2.1.16 Heat Treating of Metal Fasteners

Heat treating is often a post-forming process step in the manufacture of metal fasteners, especially those made from steel. Heat treating is an important specification on the engineering drawings for metal fasteners. Unfortunately, heat treatment is not always understood or monitored as well as it could be to reduce variability in fastener performance potential.

Heat treating is the process that carefully develops a fastener's tensile, yield, impact, and fatigue strength as well as determining wear and indention properties. Many ferrous, or iron-based, fastener materials, such as steel, are designed and intended for a specific load range, a specific temperature, and a specific fatigue strength.

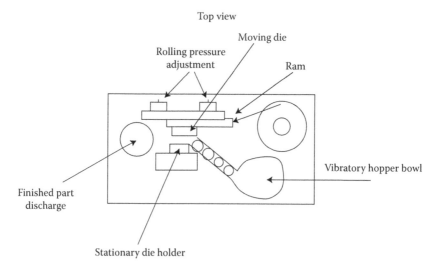

FIGURE 2.13
Production roll threader.

Ferrous-based fasteners are heat-treated:

1. To relieve internal stresses
2. To refine grain size or produce uniform grain throughout a fastener
3. To change microstructure
4. To alter the surface chemistry by adding or deleting elements

Heat-treating processes take place through a controlled combination of time, temperature, and furnace "atmosphere." For many fastener applications, steel heat treatment can be divided into through or neutral hardening and case hardening. Both transform the fastener material from the workable microstructure needed to manufacture the fasteners, either spheroidal structure for cold heading or laminar for machining, to martensitic for load carrying as a useable fastener.

A convenient way to picture fastener heat treating is as a three-step process. The three steps are harden, quench, and temper. They can be done continuously, which is the most efficient process for high-volume fastener heat treating. They can also be performed in batch steps. Or as a combination, such as harden and quench sequentially with tempering as a separate batch process.

To understand what occurs in the harden, quench, and temper steps of fastener heat treating, it is important to know something about the fastener's metallurgical structure. Steel is a mixture, or alloy, of mostly iron with carbon, manganese, and other elements added as specified for each alloy type. Many countries, including the United States with its American Iron and Steel Institute (AISI) steel numbering system, specify steel alloy chemistry with a numeric designation such as A 4037 or C 1022. More comprehensive UNS (for Unified Numbering System) metal numbering systems have also been established.

Specific concentrations of each element in the fastener respond to the temperature and atmosphere in the heat-treating furnace. Normally, either case hardening of low-carbon fastener steels or through (also known as neutral) hardening of medium carbon and alloy steels is performed. In the first step, the furnace temperature is at a level that raises the fasteners to a temperature over their lower critical temperature. This temperature can be found on the specific material's time–temperature–transformation (TTT) chart. For 4037 alloy, the hardening temperature would be approximately 1575–1600°F. Although moving in a continuous, belt-type furnace, the fasteners would be allowed sufficient time to "soak" at temperature. As they do, the fastener material's molecular energy rises and a structure change occurs. The grains of steel transform and the individual molecules of fastener alloy elements are free to form new microstructures. Figure 2.14 shows an example of a fastener's TTT chart.

Once at temperature, the fasteners are next rapidly cooled, once again using the time indicated by the fastener material's chart as a reference. Cooling within the material's transformation time yields a new microstructure,

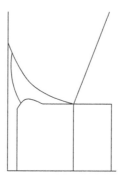

FIGURE 2.14
TTT chart—time, temperature, transformation.

higher hardness, and strength. If the blanks going into the hardening furnace were cold headed and had a spheroidal microstructure, coming out of the quench tank, they will now be mostly martensitic in microstructure. I say mostly because while at 1600°, this elevated temperature resulted in a loose microstructure of austenite. On cooling in the quench tank, the atomic elements "kick over" to the tough martensitic structure due to the rapid heat energy transference from hot fastener blank to relatively cool (180°F) circulating quench oil. As can be envisioned, not every fastener in the heat-treating lot, or every area in individual blanks, transfers its heat at the same time or rate. Those that transfer heat slower than the TTT chart quench rate remain frozen in the austenitic microstructure. The measure of an optimal harden and quench rate is a very low level of this retained austenite. It is possible with additional cryogenic treating to transform even small amounts of retained austenite to the more desired martensite. As can be realized, this rapid quenching develops high stresses in addition to hardness, so the third and final step, tempering, draws the fasteners back to the final specification hardness while relieving some of the quench stresses. If we wanted a nominal HRc 40 hardness, we would temper at 725–750°F approximately. For a higher hardness, we would reduce the tempering temperature and for a lower hardness we would increase it. For example, if for our same 4037 alloy steel fastener, we specified a hardness of HRc 26–30, and a tensile strength of 125,000 to 145,000 psi, we would increase tempering temperature to around 1100°F.

When the fastener steel is of a low-carbon type, such as C-1018, the furnace would normally be set up to add carbon to the surface of the fastener lot being put through the harden, quench, and temper process. In this setup, a hydrocarbon gas, typically natural gas, is "cracked" in a gas generator breaking it down into carbon monoxide, dioxide, methane, and other constituent gases. Carbon in excess of the carbon content of the fasteners, in this case nominally 0.018% carbon, is pumped into the hardening furnace where it is absorbed in the heated surface of the fasteners and they become case hardened on quenching, with a hardened surface and a heat-treated but more ductile core. Many self-tapping screws, self-clinching hardware, and blind rivet mandrels are heat-treated using a case-hardening process.

Another heat-treating process is vacuum or "bright" hardening. It is used for heat-treatable stainless steel fasteners such as type 410. The harden, quench, and temper steps are still undertaken; however, the hardening furnace is a vacuum pressure vessel, and air is pumped out to a practical vacuum so that oxygen in the air does not oxidize and discolor the stainless surface. The quench medium is a cool, inert gas. Tempering is also used to draw the fasteners back to their specification hardness, relieve quench stresses, and impart ductility to the heat-treated fasteners.

It is worthwhile to note that 300 series stainless-steel fasteners such as those manufactured from the 18-8 series type 302HQ (UNS30430) cannot be hardened by normal means other than from the "cold working" during manufacturing, but they can be annealed in a heat-treating furnace by heating them to a temperature of 1950–2000°F and then cooling. Figure 2.15 shows a continuous fastener heat-treating furnace for either carburizing or through hardening.

2.1.17 Checking Fastener Hardness

In checking the hardness of a specific type of metal fastener, some general testing guidelines can increase the reliability of your readings. Always use a hardness tester with a known calibration that is up to date. It is good practice to make an initial set of readings on a test block for verification of tester accuracy and to determine whether the tester error is on the high or low side of the hardness etched on the test block.

Next, in positioning the fastener, or fastener section to be tested, remember that the hardness tester operates by measuring how far the indentor travels into the fastener to be tested. If movement of the fastener testing surfaces occurs, the hardness tester still indicates this as indentor movement and a lower hardness. The fastener should have two flat, parallel surfaces. If the fastener to be hardness tested does not have two such surfaces, carefully polish, grind, or machine two appropriate surfaces for testing use. In positioning the indentor on the testing surface, avoid edges or other

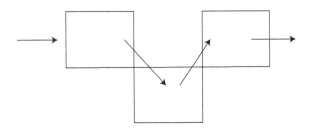

FIGURE 2.15
Continuous fastener heat-treating process.

thinly supported areas, which may yield under the indentor's applied testing load and give a false low reading. It is good practice to discard the first
reading as a setting of tester mechanism and then take a series of readings,
in the same surface area but with enough space between indentor locations
to prevent displaced material from one indentor biasing the next hardness
reading.

Standard fastener hardness testers can be either a Rockwell standard tester, with C and B scales, the most frequently used for fasteners, or Superficial
Rockwell scale hardness testers for smaller fasteners using N and T scales.
For more in-depth hardness testing, a microhardness tester using a mounted
and sectioned fastener specimen often yields the most accurate indications
of both fastener surface and core hardness as well as fastener hardness gradients. Be sure any plating or coating is removed before testing. Figure 2.16
shows a typical fastener hardness tester.

2.1.18 Screw Machining

Long before metal wire and rod were cold forged into large volumes of industrial fasteners, cutting tools were turning metal bars into precision screws,
nuts, pins, and other fasteners and assembly components. Although the
cold-heading process of manufacturing fasteners stresses the raw material
beyond its yield strength but below its tensile strength, the screw machining
manufacturing process produces the fastener by removing all excess stock
by cutting. The raw material starts either slightly above or at the largest
diameter or across the flats dimension of the finished fastener.

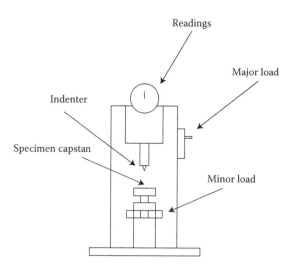

FIGURE 2.16
Fastener hardness tester.

The bar stock generally has a grain structure that is laminar, in that it runs longitudinally through the bar, which in many cases is 8 or 12 ft in length.

In automatic screw machining, the bars are spun by the machine at the rotational speed chosen for the setup. Cutting tool surface feet per minute is a process control factor with the correct material surface speed resulting in optimum chip load at generation. A constraint is that surface speed increases as the raw material is turned down in diameter. The cutting tool effectively shears the excess stock away from the final fastener geometry, with the excess discharged as chips. For more expensive raw materials, some of the chip material cost is recovered by selling them for recycling.

The cutting tools are either single point or form tool in geometry. They can either be driven by a lead screw timed to the raw material rotation, as in the generation of a screw thread helix, or off of cross slides, turrets, or other screw machine accessory attachments. By stopping and locking the raw material collet, slots and milled features can be machined into the momentarily stationary blank.

Although the screw machining process is slower in its production rate of fasteners and assembly parts, it is capable of tolerances, finishes, and features that sometimes are not producible by other processes. Figure 2.17 shows a screw machine used for the precision-turned fasteners.

2.1.19 Other Tension Fasteners

While not all-inclusive, the following designs are also often applied in tension-fastening applications:

- Threaded rod
- Eye and J bolts

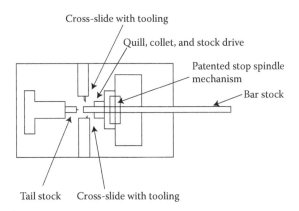

FIGURE 2.17
Screw machine used for precision-turned fasteners.

- Double-ended studs
- Self-drilling cap screws

It is a good practice to make an initial classification of fastener service loads and candidate fasteners by type using the tension, compression, and shear categories. This can help avoid using a fastener inefficiently.

2.2 Compression Fasteners

2.2.1 Set Screws

Set screws are normally headless threaded fasteners, which shorten under preload, as opposed to tension fasteners which elongate under load. Common torque-transmitting features are hex sockets and square protrusions above the threaded section of the set screw. They can be manufactured from a range of materials; however, alloy steel, or 300 series stainless is typical for the ANSI standard B18.3 hex socket screws, which are the most common. An application advantage of the internal hex set screw is its relative freedom from catching loose objects it comes in proximity with when installed in rotating machinery. At its inception, this design of compression fastener, often used to secure collar-type devices to shafting, was called a "safety" set screw due to this recessed feature.

Set screws have a formed or machined point that can locate or push into a softer mating part. While a simple cup-shaped point is most common, external cone points, half, and full dog points as well as flat and oval points are in the ANSI standards and can be used to provide application advantages.

Inch System		
Nominal Diameter	Hex Socket Size	Min. Socket Engagement
#0	0.028	0.050
#2	0.035	0.060
#4	0.070	0.070
#6	1/16	0.080
#8	5/64	0.090
#10	3/32	0.10

Figure 2.18 shows a typical hex socket set screw.

2.2.2 Washers

Washers are also compression fasteners. They distribute the clamp load over a larger bearing area and perform in a "snowshoe" manner to lower

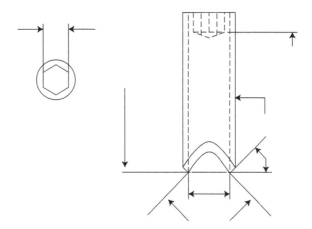

FIGURE 2.18
Hex socket set screw.

bearing stresses. A good practice is to use well-made washers in conditions where indention is a possibility. That can also increase clamp load for a given torque by reducing underhead friction during torquing. They also increase clamped grip length, which can be a performance advantage. Potential draw-backs are the increase in parts count and an increase in joint faces, which can bring on vibrational loosening and other problems. Locking washers can be a benefit for loosening. Their clamp load increase, as in the case of simple split lock washers, is limited to the load it takes to compress them. When compared to the clamping preload in the fastener itself, this can be relatively small. Belleville washers are more highly engineered and efficient in this regard. An important fastener concept is the "clamp-affected zone" (CAZ) concept. CAZ is the volume of material over which the fastener's clamping load is applied. Its shape is a function of the relative ductility of the assembly members, the size of the clamping load, and the washer area of the fastener, be it a separate component or integral to the fastener. A summary of washer functions would be as follows:

- Bearing area load carrying
- Control of the clamped assembly zone
- Control of underhead friction
- Increase in fastened grip length
- Protection of clamped member surfaces

By combining the washer as a part of a fastener head design, some of the washer functions may be taken care of; however, it is good fastener engineering to consider these washer functions when designing fastened assemblies. Figure 2.19 shows a typical washer used in general assembly applications.

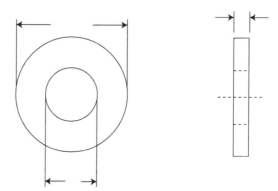

FIGURE 2.19
Typical industrial washer.

2.3 Shear Fasteners

Shear fasteners are those designed to carry loads perpendicular to their axis. Good examples of shear fasteners are dowel pins and roll pins. Dowel pins were an early form of fastening to carry shear loads and are still a very efficient assembly method. Dowel pins can range from simple low-carbon wire cutoffs to alloy steel, and hardened and ground tool-and-die quality dowels with end chamfers and radii. While dowel pins are manufactured from solid round stock, roll pins are fabricated from strip steel, which is coiled and cut into pin shape with springy characteristics. In general, dowel pins provide much greater shear strength, while roll pins provide assembly economies of installation.

As an application note, dowel pins should be pressed into the assembly, not hammered, as is sometimes done. This impact, especially on hardened dowels, can present a shattering hazard.

Dowel and roll pins are available in fraction sizes, usually from 1/8 in. diameter through 3/8 in. in 1/16 increments, and common lengths. A typical list of fractional sizes would be 1/8, 3/16, 1/4, 5/16, and 3/8.

2.4 Adhesives

Adhesives for fasteners cover a wide range of types. Generally, adhesives for fasteners can be used as stand-alone fastening systems or as supplements to mechanical fasteners. When used as supplements, they can be specified based on the relative strength and permanence. An adhesive listing would include:

Screw grade	Low strength, for up to 1/4 fasteners
Nut grade	Medium grade, for fasteners 1/4 and over
Stud grade	High strength, good for grades 5 and 8

Adhesives can either be specified preapplied or applied on site. In either case, the absence of air when installed allows them to cure.

2.5 Fastener Manufacturing Methods

A lot has transpired since I wrote the first edition of *Mechanical Fastening, Joining, and Assembly*. Nearly 15 years have gone by, and during this time, fastener manufacturing and use has ebbed, especially in the United States. More recently, fasteners and fastening are experiencing a most welcomed resurgence. As I read the preface to the first edition these many years later, the thoughts I expressed then about the rise of computation and consumer electronics skill and usage combined with a decline in basic mechanical knowledge and expertise rings even more truly today. Then, mobile phones were large and uncommon. Today's mobile communication devices (one really cannot call them just phones) and portable tablets place the information and power of the Internet available to many, worldwide—for good or not so good purposes.

It has become popular to speak of manufacturing in North America in public discourse. This is in my view definitely a good purpose. In this chapter, we will expand on the many fastener options and procedures available to this new generation of manufacturers. We will also further expand the reader's understanding of the cold-heading and thread-rolling fastener manufacturing processes along with the tooling designs and materials used. These are fundamental to the manufacture of a significant portion of industrial fastener production and important fastener manufacturing processes. It can also provide guidance on what is possible if the interest or application develops requiring other than a standard fastener design. Large quantities of these engineered, special fasteners can be found in automobiles, aircraft, consumer appliances, and numerous other fastened applications.

We can divide the general shapes possible with cold heading into three categories. These are upsetting, or heading, forward extrusion, and backward extrusion. We will explain each and provide examples of the cold-header tooling, which might be commonly used for these three operations.

Starting with upsetting, this cold-heading process forms the straight wire section, usually above the face of the cold-heading die, into a head. A very simple upset familiar to all is the head of a nail. The nail is cold headed from wire with the head formed on a cold header. To understand how this works, let us consider a column of steel. Let us make the wire diameter 2 mm. We can have a very short column of just a few millimeters or we could have one considerably longer. In cold-heading terminology, we can use an index of diameters to express column length to diameter. A 6-mm length would be 3 diameters of our 2-mm diameter wire. We know from practical experience that as the length-to-diameter ratio increases, the column stability decreases. A short, wide steel column is stable, whereas a long thin one is less stable. This proportionality is known as the slenderness ratio. It is important in building columns, and also in cold heading. Referring to the simple single die, two-blow header in Figure 2.20, a column of cold-heading steel wire of approximately 2 mm diameter can be struck on its end and formed into the head of a nail, or many other useful fastener shapes. This chapter shows just a few of the almost infinitely possible shapes. The simple cold-heading die and punch set as sketched in Figure 2.20 shows what this simple heading tooling might look like.

The rules for upsetting lay a good foundation that the user should know. The wire length protruding above the face of the cold-heading die as the punch is delivering its cold-heading blow most not be above a specific slenderness ratio for a given material formability—if it is, it will buckle rather than cold flow into the desired fastener head shape as defined by the cold-header tooling's punch and die. The cold-heading rules for upsetting are 2¼ diameters of material can be upset for each blow. Multiplying 2.25 × 2 mm gives us 4.5 mm, and multiplying 4.5 × 2 mm yields 9 mm of cold-heading stock protruding out of the face of the heading die when struck. This volume of upset material then defines the volume of material in square inches or millimeters that can be upset in our single die, two-blow, cold-heading machine. In observing the various head styles of tension fasteners in this chapter and in your own applications, the volume of material in the head is directly related to this upset column if it is a cold-headed fastener. The volume is calculated as the square of the head diameter times the head height. Most common screw head styles have a proportionality of head diameter and head height to be upset well within the 4.5 diameter limit. If the head height

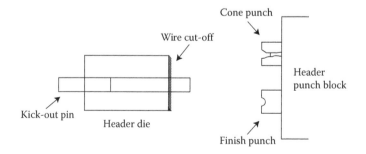

FIGURE 2.20
Header die, simple.

is lower, say on a truss head style, the head diameter can be larger. The limit is approximately five times the body diameter. For a 2 mm body diameter, we would expect even a truss head or similar "special" to be less than 10 mm in diameter. In terms of the basic cold-header tooling, a straight hole die, solid punches, a die know pin, and a cutter and quill comprise a set of header tools. The sketch in Figure 2.20 shows solid punches. Notice that the second punch is the mirror image of the final shape of the upset head. The first punch is where the art of cold-header tool design exerts its influence. The first punch, often called the cone blow, is the preforming operation that starts the upset and moves the fastener raw material into a shape that is most conducive to a clean finishing blow by the second punch. The second punch must be the shape of the desired fastener head. The first-blow punch is often called the cone blow because if we could stop the cold header between the first and second blows, we would see a cone-type shape. For each head style, a specific cone shape is required for optimum material flow. Referring to Figure 2.21, upsetting is a process which requires force sufficient to move the upset material into the yield zone of the stress–strain diagram shown. However, additional stresses of the material will cause it to split or fracture. This is the reason small discontinuities can often be seen on the edges of even common fastener shapes such as truss heads. The material traveling between punch and die faces, already into the yield zone in those areas which crack or split are stressed past the fastener raw material's ultimate tensile strength. Adding to the skill in tool design required is that fact the fastener cold-heading wire cold works as it is moving. Some materials, such as many grades of stainless steel, have work hardening rates, as shown in Figure 2.21, which decrease the amount of forming possible before over-tensile cracking occurs. Finishing our initial review of cold-heading tools for

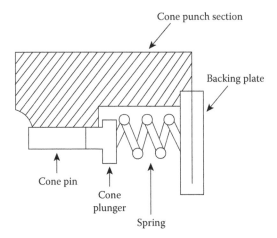

FIGURE 2.21
Sliding first blow punch.

this simple single die two-blow fastener blank forming process, we would
have several other pieces of tooling. Wire tools include straightening and
feed rolls, a quill and cutter, cutter finger and cutoff transfer, and a knock-
out pin. If the wire is being drawn to size at the header, which can be advan-
tageous, a draw die will also be used. Going through them starting with the
wire tools, these rolls contact the wire and pull it from the coil. Header coils
can be anywhere from tens of kilograms up to hundreds. The straightening
rolls contact the wire to remove the curvature of the wire resulting from it
being wound in a coil. A typical set up would have a series of opposing rolls
in the X–Y-planes. Following the wire straightening rolls are the feed–rolls.
These are larger in diameter and driven by the cold header with rotation
timed to the cutoff, 1st blow, 2nd blow, kick out of the header. If wire is being
drawn at the header, the drawn die will be in front of the straightening rolls.
As the feed rolls advance, they apply sufficient force with the outside diameters of the feed rolls to advance the heading wire to the distance required to
form the blank that is being headed. The volume of this blank in cubic millimeters must equal the volume in the entire blank, from upset head to point
of the shank. If a wire draw die is used, it is powered separately by a wire
drawing machine with some form of buffering with a draw die drum to
avoid having to time the wire drawing speed to the feed roll speed. The feed
roll outside diameters will have a radius or "vee," which will match the wire
diameter being cold headed. The wire feed roll force must be enough to feed
the wire through the quill until it hits the wire stop. Not enough pressure by
the feed rolls on the wire and it will slip; too much pressure and the wire
could be deformed or marked. The quill will be just large enough at its face
to provide a good cutting edge. Carbide is often used as an insert for quills
and draw dies, whereas hardened tool steels are used for the other wire
tools. The quill may have a bell move shaped entry and the wire may also be
guided by a feed tube between the feed rolls and the back end of the quill.
Once the wire is driven by the feed rolls through the face of the quill and is
met by the wire stop, the feed rolls may be timed to slide just a slight number
of degrees to hold the wire firmly. Cold headers are rated by the wire size
they can cut off, normally based on a low-carbon steel material. They also
state the length of wire that can be ejected, or "kicked out of" the cold-heading die. Fasteners that are either very short or very long in the range of wire
cutoff length and die kick out required can present some cold-header tooling
and tool setup and timing challenges. This is how we could determine the
maximum length part kick out plus upset volume, along with the maximum
cutoff length. The feed roll circumference and maximum degrees of rotation
per feed stroke cycle of the header crank would provide the maximum length
cutoff length. The kick-out stroke at its maximum back position in the header
die would provide room in the die bore for the longest shank possible for
that cold header. A kick-out adjustment device is usually found at the back of
the cold header to adjust the kick-out length and time its motion, usually
with a cam connected through gearing with the main crack shaft. The

kick-out length is less than the cutoff length. The difference in length is the stock which will be protruding in front of the face of the die when the knock-out pin comes to its adjusted length stop. As the first-blow punch impacts the upset end, it will flow into the cone blow shape if we have used the proper header tooling and performed the correct machine setup and timing for this job. If we want to upset a head of greater volume, the shank length will need to be reduced on this cold header. In fact, during set up, it is common practice in my experience to under-fill the upset slightly to try out the tooling and setup. Once setup was confirmed, additional stock was added to the upset to completely fill at the head to specification dimensions. This was done by either decreasing the knock-out stroke, increasing the feed length, or a combination of the two. It is hoped this provides some insights into the simple upsetting process in fastener cold heading. Before we move on to forward and backward extrusion, it would be useful to provide the reader with an idea of the creativity that can be used in the tooling of even a relatively simple, single die, two-blow header. Figure 2.22 shows some more advanced tooling that can be used to produce standard and special fastener and cold-headed components with more complex geometries. The first piece of cold-header tooling is a sliding first blow, shown in Figure 2.23. Remember that the first blow strikes the end of the wire to be upset at speed and with impact. Often, the end will not be completely smooth, square ended, and symmetrical. It is common to put a 5 degree angle on the face of the wire stop to provide a reaction to the cross cutting motion of the cutter as it shears the wire from the coil end at the face of the quill. The cut end left in the quill is then pushed forward by the feed rolls and is itself then cut and positioned in-line with the cold-heading die for loading in, heading, and kick out. This first-blow impact is often controlled through the use of a sliding, or telescopic first punch. They can vary in complexity. The sliding first-blow punch in Figure 2.23a uses a punch case, spring-loaded heading pin, and machined and adjusting fastener-provided provision in the punch block for the punch to compress. As the punch block approaches the face of the die, at speeds of

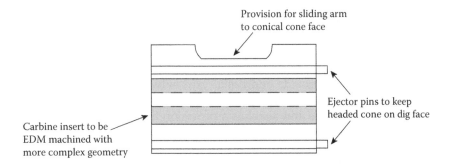

FIGURE 2.22
Special first blow punch.

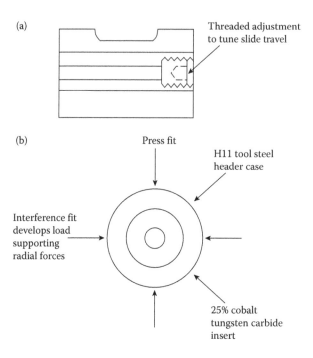

FIGURE 2.23
Sliding blow punch.

7000–13,000 pieces per minute in my experience, the cone shape opening covers the wire end. The wire is pushed to load into the heading die, pushing the knock pin back to its stop. Once the knock-out stop is contacted, the upset end starts compressing the punch spring until it is coil bound. The head blow then flows the material into the cone shape of the punch before moving away from the die face, leaving an upset first-blow cone. The expanding spring and heading pin can be made to help keep this shape in the die. A work piece that pulls out of the die can cause a tooling smash-up, which at the least is expensive and time wasting on the fastener production floor. Often, the cutter, sliding first blow, and timed knock out are all designed to work together in a synchronized, high-speed forming of wire stock to fastener blank. If the material being upset is so large and/or the shank in the die is so short as to still cause mis-feeds, annular rings have to be machined into the bores of dies to provide additional grip between the first and second blows. The punch block itself has adjustments to center the punches on the heading die center line and also has a provision to shift the punch so after the first blow, the second blow punch is transferred into heading position with its centerline coaxial with the heading die. The heading die itself can be a sold piece but is often built of a high-speed steel case, with H13 being a very common cold header case material. The carbide insert is more expensive to manufacture and machine. Its benefit is a greatly increased wear resistance at heading speed. Figure 2.23b shows a header die with tool steel case and carbide insert. The outside of the case has a surface for the point of the header block locking bolt to contact to lock the header die into the die block. As can be visualized, the mouth of the heading die will become bell mouthed with wear, especially if care is not taken to assure the wire is feeding centrally into the die from the cutter. The stock riding one edge of the header die can be a common problem. In addition to decreasing die wear, the resulting fastener blanks can have undesirable underhead radii, which can cause problems when attempting to thread-roll blanks with a bell mouth condition. So far the header dies have been of a through-bore design. The resulting fastener blanks from these dies will be straight and round. If it is to be a rivet, the shank diameter will be our rivet diameter. If it is to be threaded, the cold formed shank should be at the required roll-thread diameter, which is typically close to, but not precisely the same as the pitch diameter of the intended screw thread. There are many fasteners that have more complex shank geometry requirements. A typical one is a header point. These are desirable to give the rolled threads a slightly smaller starting thread diameter. Figure 2.24 shows a typical header point cold-heading die. Header point angles are typically around 15° per side. The wire is extruded forward into the header point section of the die. The knock-out pin will usually be smaller and needs to carry higher knock-out stresses. Many fastener shapes have shoulders or heads that are made inside the face of the cold-header die. Figure 2.25 shows a shoulder-type blank die design in view (a) and a head that will be formed at least partly in the cap, which is pressed into the face of the die in view (b). Many screws with sockets such as hex or other internal drives are made in a cap, usually from a high impact capable tool steel pressed into the front of the cold-heading die. These are just a few of the many upsets that can be cold headed for standard and special fastener shapes. Creativity and the need for other geometries can make many fastener shapes candidates for cold heading. The latch that can be found protruding from the sides of doors and offices worldwide is now cold headed. It is "D" shaped and cold upset from round wire drawn into a "D" shape. The other operations which are key benefits of the cold headers used in fastener

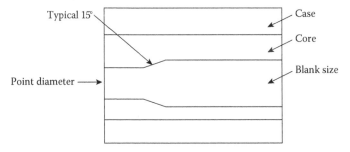

FIGURE 2.24
Forward extrusion die cross section.

manufacturing are forward and backward extrusion. Extrusion features on a cold-headed fastener blank can be either along the shank down to the point and up in the upset head. In fact, the header point we just discussed is a form of forward extrusion. As its name suggests, forward extrusion moves the raw material forward as it reduces it in diameter. The reduction in diameter is a key factor in the limits of forward extrusion on a cold header. The reduction of area varies with the degree of formability of the raw material. Low-carbon steel has formability, while alloys steels, stainless steels, and higher nickel content materials as well as the non-ferrous alloys have a range of extrusion capabilities. We rate the forward extrusion by the reduction of area. The percent reduction in area is obtained by dividing the new circular area by the starting diameter's area. For open extrusion, the limit of reduction of area is approximately 35%. In open extrusion, the material is entering the cold-heading die and is being pushed from outside the face of the die into the die bore and extruded through the extrusion land of the die. Figure 2.25 shows an open extrusion heading die and points out the extrusion land, extrusion angle, and extrusion relief. Some frequent fastener blank cold-headed extrusion features are the roll-thread diameters of screw with grip lengths at screw thread major diameter or larger. Moreover, dog-type points and round forward extrusions, known as tenons, used as locating points are also typical of forward extrusion. When reductions in area of greater than those feasible with open forward extrusion are required, closed trap extrusion is available. In trap extrusion, the material is completely trapped in the die bore. This allows for much higher forward extrusion pressure to be

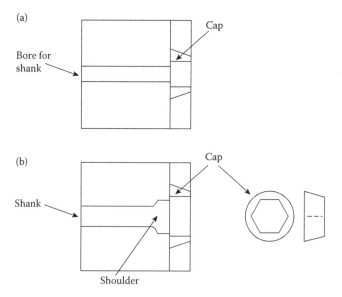

FIGURE 2.25
Open extrusion die cross section.

applied to the end of the blank. The dies and cold-heading machine must be capable of delivering and withstanding these increased forces. Figure 2.26 shows a trap extrusion die. Many trap extrusions are done on a type of cold header designed to include this operation. The two-die, three-blow header became widely used for this type of fastener work. The first die is typically used as a trap extrusion station on receiving the wire cutoff. The resulting trap extruded preform is kicked out by way of a knock-out pin or device into a set of transfer fingers to the second die where a cone and finishing blow can be delivered to complete the cold-forming operation of the blank. Shoulder bolts as found on many automobiles are textbook trap extruded two-die, three-blow cold-headed fasteners. In back extruding, the material is made to flow back toward the punch block as the name suggests. Hex sockets, as well as many heads with internal forms, are examples of back extrusion. Once the material is confined by the die, it provides a reaction to the forward motion of the punch. As that reaction is overcome, the material yields and starts to flow into the die and around the punch. If the tool designer provides clearance around the punch, say a hexagonal shape, the material will flow back along the hex punch. This is back extrusion. If our back extrusion fastener blank is to look acceptable to standards, the tooling clearances must be well designed, set up, and operating well. Back extrusion is also used for socket wrench extension bars. This is an application I had considerable experience with early in my career. It gives us a great example of back extrusion and upset of the head and forward extrusion of the point end square drive. Continued development of the single die and two-blow cold-header tooling advanced it to adding a forward-extruded external square drive on the point end of the cold-headed extension bar blanks. This replaced the former drop-forged extension bars with drill and broached sockets and broached square drive ends. Key factors to understanding and using forward and backward extrusion on cold-headed fasteners and special components are the limits any material will flow forward or backward into the extrusion tooling and also material-tooling frictional drag. Recall that the cold-forming operating range involves pushing the material past its yield

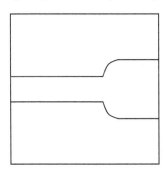

FIGURE 2.26
Trap extrusion die.

point but not past its ultimate tensile strength. Due to the small clearances, cold-headed extrusions can exert extreme forming pressures. Additionally, there are the surface areas that develop significant frictional forces, which further increase material and tooling stresses. With respect to part design, extrusion angles and radii can assist in helping the material flow into the required extrusion shape. Cold-header lubricants have also been developed with extreme pressure properties to reduce friction. And tooling design can help. Some examples are sliding punches and dies, which can alter the shape of the forming forces so that it is applied over slightly more time during the cold-forming cycle, which can drop stresses and friction to limits that allow satisfactory manufacturing and part net shape. Two good examples of cold-headed fastener backward and forward extrusion are the 12-point cap screw and the semi-tubular rivet. The 12-point head, with integral flange washer, is back extruded into a 12-point shaped internal punch with a pin sliding inside to help deliver the forming forces to flow the material into the 12 points and then to keep the blank in the die after the blow, usually with a punch spring or more commonly a timed punch knock-out device. Twelve-point heads are found in higher performance cap screws, often formed from more expensive raw materials. At the other end of the cost structure, semi-tubular rivets have circular point sockets, which present the same forming and pullout technical issues of high forming forces and forming friction, but on the forward extruding side. Again, special extrusion pins and stripping mechanisms in the cold-header tooling and machine solve these in production. A sketch of a semi-tubular rivet extrusion pin is shown in Figure 2.26. Note that the side of the forward extrusion pin is relieved, meaning the material does not create frictional drag along the entire length of the extrusion.

It is hoped that this gives a more detailed understanding of the cold-heading process used so widely in the fastener industry. Many other cold-forming processes exist. Trimming can be used to shear flash from the edges of surfaces such as hex flats. Piercing can be used to punch holes through parts such as hex nuts and other nut-type cold-formed blanks. The range of cold headers includes not just the two types but 2-die, 2-blow part formers with 4 and 5 dies and blows and bolt formers with integral thread-rolling capabilities. That said, the concepts of upsetting and forward and backward extruding help in understanding the fastener cold-forming process. When some heat, usually through induction coils, is added, it becomes warm heading; although the term cold heading is in some ways a misnomer in that the fastener blanks discharging from the machine are hot to the touch, and care and hand protection is required in handling them. The microstructure of the metallurgy does not change although grain flow can be improved. Additional heat takes the process to hot heading. In hot heading, the metallurgical structure is transformed. Hot forming, or forging, is often the most efficient fastener manufacturing process for some materials, part geometries, and lot sizes. Some additional cold, warm, and hot-forged fastener types are shown in the shapes in Figure 2.27.

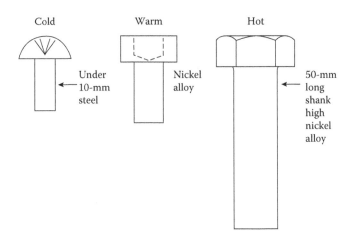

Cold   Warm   Hot

Under 10-mm steel

Nickel alloy

50-mm long shank high nickel alloy

FIGURE 2.27
Heading.

2.5.1 Roll-Threading and Other Screw-Threading Processes

Many years have passed since the first edition material was written. While flat die roll threading, as shown in the top view of the thread roller shown in Figure 2.13, remains a dominant fastener manufacturing process, it is one of the numerous screw thread manufacturing processes we will explore. A noteworthy development since the first edition is the change, not necessarily for the better in terms of quality, which has taken place in global manufacturing. Several recent incidents of lead errors in screw threads whose root cause was lead error in the threading dies have occurred. With threading dies being made in sources spread far and wide around the globe, threading tool quality has varied more in recent years.

Thread rolling is a forming process and is one of the more widely used screw thread manufacturing processes. It is not by any means the only one used in fastener manufacturing. We will update our review of them. Flat die thread rolling, as illustrated in Figure 2.13 is still the predominant screw thread manufacturing process in fastener factories worldwide. Many of the flat die thread-rolling machines are quite old. If well maintained, they can continue to produce quality screw thread with respect to applicable specifications and thread form control. The following text describes a set of incidents indicating the changes that have occurred since the writing of the first edition. A well-established socket fastener products company was threaded rolling an important motor vehicle shoulder screw with metric screw threads. As was this company's practice for many years, the screw threads were measured in process using variables gaging with pitch diameter rolls and functional segments, which were set to a master setting plug. We will discuss this type of systems, 22 screw thread gaging as part of a larger section of screw thread inspection and control, in a later chapter. Suffice it to say that the screw thread being rolled were measured less precisely, as indicated by the difference in indicated size readings between pitch diameter and functional size. Records kept for this part over many production runs of this fastener showed that the new setup was producing screw threads that, while within the technical specification for this metric thread, contained thread form that varied from the quality that the motor vehicle manufacturer and the users of their vehicles relied upon from this critical fastener. After eliminating all other setup and blank quality variables, someone decided to try rolling the screw threads with another manufacturer's pair of thread dies of the same size. When the new flat dies were set up in the thread roller and blanks rolled at production speed again, the pitch diameter/functional diameter came back into the normal good quality range. The variance, which happened to be lead error, was in the thread-rolling dies. These were supplied by a fastener tooling manufacturer with decades of experience supplying these type of flat dies to this fastener manufacturer. Further investigation for the root cause showed that these threaded dies had recently been outsourced to an offshore source to reduce cost. Within a short period, another customer experienced a similar lead error in threading dies and took corrective action. A question to be asked is how many other sets were used and the difference not noticed? Flat die thread rolling has the advantages of high speed. Thread-rolling speeds of 400–500 ppm are not uncommon. One small-diameter flat die thread roller was nicknamed the hummingbird and can roll at in excess of 100 ppm. It is found, however, that the sliding nature of the thread-rolling surface of the blank on the die faces will overheat and burn up the thread die form, making less than full speed thread rolling usually the most efficient thread-rolling speed with respect to parts per shift or per die setup. Cylindrical and planetary thread rolling are also commonly found employed in fastener plants. They use a similar flow of roll-thread diameter into the thread die-forming process. Round threading dies have a benefit in controlling out of roundness in the product thread being produced. Two flat dies roll threading a round blank typically cannot completely eliminate two point out of roundness. A design experiment conducted by the author using a #4–40 UNC-2A screw thread blank and a Hartford horizontal flat die thread roller could reduce out of roundness of the threads to no less than 0.0004". While not large, this is a bias of flat die thread rolling. In cylindrical thread rolling, the dies can be either two and at 180° or three and at 120°. The work pieces can be presented to the thread-rolling dies either horizontally or vertically, depending on the machine. Radial thread forming pressure is reduced as is the wear rate on the thread-rolling dies. The dies are, however, much more expensive to purchase. Shorter production runs, tighter form control, and the ability to thread roll more difficult shapes such as thin-walled parts, or very long thread lengths are advantages. A special form of cylindrical threading die machines is the planetary thread roller. In this process, an inner thread ring and an outer die "shoe" roll thread the rotating blanks, several at a time.

The inner thread ring rotates at relatively lower speeds compared to the reciprocating rate of an equivalent sized flat die machine, but parts can be inserted between the dies at a much higher rate. The benefits of this are a very high thread-rolling production rate and lower die wear. Thread-rolling attachments are also used frequently in the production of fastener products on screw machine and CNC machine equipment.

Thread milling has gained in usage in CNC machined external and internal threaded components. It has specific advantages when screw threads are required on difficult-to-machine work materials such as high nickel content and aerospace alloys. An important quality control factor is the relative diameter of the thread mill and the screw thread nominal diameter. The author has seen screw thread form variance in an application where a thread milling cutter was used for several diameters having the same pitch without regard to the relative curvature of mill diameter and product pitch diameter. A final screw threading advancement has been thread whirling. In thread whirling, a thread milling cutter is again used. The difference from thread milling, where the mill both spins and rotates around the screw thread circumference, is that the work piece rotates counterclockwise to the thread mill's rotation. Rotating speeds and tool/work piece positioning are key factors in the quality of the thread forms whirled. Good results have been obtained with thread whirling on difficult-to-machine threading work. As we leave this review of threading, a brief mention of two traditional standbys: thread tapping on internal screw threads and thread chasing of external threads still have a place in specific fastener applications. In both, the proper-sized work diameter and tap or die size as well as tool speed, rigidity, and proper cutting lubrication, in addition to tool sharpness and operator technique are key factors. Having personnel new to industry trained in these basic threading processes is a sound starting point for a better understanding of fastener manufacturing processes.

2.5.2 Some Additional Thoughts on Adhesives

Much has happened in the field of adhesive assembly in the recent generations of new products. Automatic application systems combined with a better understanding and control of assembly surface preparation has increased the reliability of the resulting assemblies. For the consumer, division of adhesives into two-part epoxies, anaerobic adhesives, and organic mixtures provides a wide range of options. In building construction, the factor of outgassing of assembled structures is an area that may warrant better understanding. One very instructive traditional structural adhesive application is the glue used to fasten the top, back, and sides of stringed musical instruments. Of particular interest due to the range of build quantities are acoustic guitars. Some acoustic guitars represent a significant monetary investment with respect to both financial cost as well as labor and raw material investment. Woods with desirable sound-producing qualities can be costly. The skilled labor to design and assemble an instrument that will project music that is pleasing to the ear in the hands of a skilled musician represents design and manufacturing that is unique, combining demands that can rival many seemingly higher technology products. It also has a tradition that can trace its origins back to ancient Italian violin makers. These instruments successfully used early adhesives and procedures that yield assembled joint integrity and reliability that can span decades. I had the opportunity to tour a guitar-manufacturing facility and recommend it as an understanding of adhesive assembly processes.

2.5.3 Some Additional Thoughts on Welding

Welding can also provide longevity in terms of joint reliability and integrity. Of particular interest is the automotive industries' continual evolution of automated spot welding. The ability to join complex sheet metal shapes into automobile bodies which neither rattle nor otherwise perform poorly over significant dynamic loading in service is a testament to how far this fastening technology has advanced. Of particular interest to the student of assembly is the control of the entire spot-welding process: work piece preparation and fixtures, welding head motion control, and electrical power variance. Also noteworthy is the sustainability of the material used. Welded vehicle bodies, which have reached the end of their service lives, can be recycled into new products. Another welded assembly that is noteworthy is the recreational bicycle. I have observed the welds in the one I own and enjoy. These welds are of a continuous weld bead type. The bicycle frame is comprised of tubes. The weld beads can be examined for variation and exhibit good control of weld material flow and control. Perhaps as much as any fastening technology, the welded bead pattern of this fastening process gives very rapid visual clues with respect to joint quality. As an apprentice at a naval shipyard, I also was fortunate to be on hand to observe the post-weld stress relief and x-ray metallurgical inspect that welding affords. In the next chapter, we will use the welding concept of the weld-affected zone in discussing fastener mechanics.

2.6 Fastener Dynamics

A fundamental concept in dynamics is the difference between weight and mass. In dynamics involving fasteners and assemblies, mass matters. In Chapter 1, under the fastener statics section, the concept of systems at equilibrium or rest was introduced. Also introduced were forces, vectors, and reactions. In dynamic systems, assemblies are in motion. Proceeding from rest to motion, acceleration occurs. An assembly's resistance to acceleration is mass (mass is defined as resistance to acceleration). Motion can be either linear, in a straight line path, or nonlinear. We use mass as a unit to mathematically define acceleration. In the earth's gravity, weight, which is the force of gravity, is of an equivalent magnitude to mass. Once out of the earth's gravitational field, weight changes, whereas mass remains constant. Distance is measured in meters. An engineering definition of motion is defined in meters per second. When velocity is accelerating or decelerating, it is quantified in meters per second squared. In the numerical methods section, we will introduce mathematical operations. Taking the first integral of velocity yields acceleration, which is change in velocity. This is positive if accelerating and negative if slowing down. Taking the second integral yields jerk. Jerk is the rate of change of acceleration. A few fastener applications are the flat thread-rolling dies we have discussed and ordnance fired from a weapon and flying from the barrel. In the threading dies, the moving die advances from a stop to a point where the blank can roll off the stationary die. The moving die then stops and accelerates up to speed to move back past to the home position where the dies have been set up for the threads to be in matching position. The moving die slows down to a stop and then accelerates forward again with a new blank. This type of motion is called rectilinear. The fasteners clamping the dies experience dynamic forces. The masses of these components are a factor in the magnitude of the forces. The moving die is connected to a pitman arm whose other end is connected to a rotating flywheel. The mass of the flywheel store energy in its mass. Its motion is circular. As it speeds up and slows down, it has what is known as angular motion. An important dynamic concept to measure angular motion is the radian. A circle contains 360°. It also has 2π radians. Radian measure gives a convenient notation for calculating angular motion which is found throughout fastener-manufacturing machinery and fastener applications. A value of 1 rad/s gives a scientific value for circular motion. Note also here that energy is being transferred from a prime mover such as an electric induction motor, through a gear set to turn the flywheel with angular motion. This is then transferred to the moving thread die which is in rectilinear motion. Energy is being transferred from electric motor to the fastener blank being roll threaded. The mass of each component stores and then transfers this energy to do useful work. Time is an important dynamic concept in that it defines the rate at which energy is being transferred. In dynamic systems, the knowledgeable observer can read the motions and masses under linear or angular motion change and measure the time elapsing during each cycle. In statics, we reviewed forces. The energy of force per unit time defines power. To perform work, energy is required. It is stored and transferred in mass. Work for a specific time requires a specific amount of energy. In the projectile, time change is extremely rapid. Let us visualize a 11-kg projectile with a threaded conical-shaped member screwed into its nose. When the gun powder is ignited behind the projectile, its rapidly expanding gasses accelerate the projectile at a high rate through the barrel and into the air on a trajectory. Plotting its instantaneous speed at any point in the trajectory, we would observe a wide range of flight speeds, rates of change of projectile speed, and resulting dynamics forces and reactions related to relative component mass. These are important dynamic factors for the fastener person to study and understand. Impact is a dynamic concept, which can be a factor in some fastener applications. Consider the application where a person reaches up and takes hold of a shelf to help lift themselves up from a desk chair rapidly. In an actual application, the shelf fasteners may be pulled out of the wall. This is an example of impact. The applied load is known as impulsive. The engineering calculation of sophisticated assemblies needs to evaluate the natural frequency of the assembly and its ability to absorb impulsive energy. Impulsive loads, or impact, increase the applied loads and reactions for a short timer duration. If you set a 5 kg block on a bench slowly, we expect the bench to apply a 5 kg reaction and we have equilibrium. If we drop the same 5 kg block from a height of 1 m above the desk, it accelerates from the force of gravity minus a negligible wind resistance and on contact with the bench applies a much larger load for a short duration. With strain gages we could plot these forces and attenuation with time. Dynamics in fastening, then, is an important application area in the engineering of fasteners and assemblies. The time–force plots in Figure 2.14 show a few of the many types of dynamic loading, which can be found in fastener applications.

Reference

1. *American Fastener Journal, Heat Treating Fasteners*, Powell, OH, May 1993.

3

Fastening and Joining Mechanics

3.1 Metal Joints

The engineering mechanics of fastening and joining are in some ways similar across the range of assembly work that is performed every day. Forces are applied by the fasteners or fastening systems. If the job is done right, the service loads are contained by the fastening design and installation. The diversity and individuality of assembly component materials, number of joint interfaces, and intended service conditions indicate as well the scope of possible mechanics involved in assembly work and performance. With knowledge and skill in applying fastener engineering procedures to the assembly of product components, your goals can be better reached with reliability in performance at the same time that you approach the equally important assembly goals of efficiency and economy. That this often occurs with less mathematical precision or that it requires practical simplifications to everyday time/cost constraints should not deter your efforts.

In laying out the field of fastener mechanics, it will be useful for us to separate the procedures into sections according to the assembly component materials that are being joined. A logical method for us to use is to make a distinction between metallic and nonmetallic assemblies, although I am sure many can point out that there are similarities in the springiness and "plasticity" of many all-metal assemblies, just as there are conditions of high rigidity and compressive strength in numerous engineered thermoplastic products. And this is not to mention the many hybrid assemblies that are made up of both metal and plastic components, exhibiting properties of both.

That said, it is valuable for us to make the mechanical distinctions here between joint materials for the purpose of getting across the mechanical techniques and models, which can help interested students and observers of assembly engineering situations to further their understanding and improve their assembly's performance.

As a starting point, let us pick up the metal plate example from Chapter 1 and set out the mathematical tools that are available to observe the mechanical actions that occur when metal fasteners are installed in metal assembly components which are solidly joined together.

3.1.1 Rigid Metal Joints

Let us suppose that we have two cold rolled steel components, the end cap of a pneumatic cylinder and the cylinder body itself. Our goal is to have a set of algebraic equations at our command that enable us, with some relatively high degree of accuracy, to estimate the forces and actions occurring in the assembly and use of our pneumatic end product. The reasons for our quantitative interest are many. Among these reasons are that the safe operation of the cylinder, its longevity of operation and reliability will depend on the quality of our fastening. Further, the efficiency of our cylinder manufacturing and assembly will depend on the fastening design approach we choose. To make informed decisions concerning these important factors, we need good information. And much of the information we need is numerical.

   We can consider, for the sake of mechanics, that assemblies comprised of a few, or many, metal components either have spring rates for the components that are higher than the spring rates of its fasteners or that assemblies have components that are relatively "springy" in either their material, shape, or joint interfaces, or a combination thereof, compared to the spring rate of their fasteners.

   We will classify all springy fastener/less springy assembled parts as rigid joints. We will further classify relatively rigid fastener/relatively more springy-assembled parts as flexible joints. Our analysis of each of these assemblies will be grounded by this identification. Let us first take a look at the quantitative reaction of our working material, steel, to the loads it will be carrying.

   The stress–strain diagram for steel shows a linear, elastic range below the yield point and it shows a plastic nonlinear range at stresses above the yield point. The rate or ratio of strain to stress in the region from zero external stress to near the yield range is called the modulus of elasticity. A look at the stress–strain diagram in Figure 3.1, along the elastic moduli of several common assembly materials, shows the relative elasticity of these metals.

   Now let us take a look at our cylinder assembly with its fastener installed and under load in Figure 3.2 and lay out the specifications for our application in order to perform our fastener application engineering calculations.

   Application specifications
       Cylinder working pressure = 240 psi
       Cylinder bore = 2.250 in.
       Cylinder area = $3.1416 \,(2.250 \text{ in.})^2 / 4$
       = 3.976in.^2

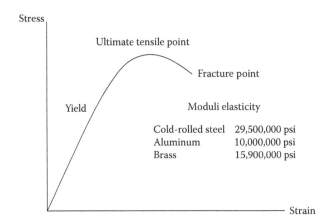

FIGURE 3.1
Stress–strain diagram for steel.

Load on cylinder head

Stress = Load/Area

Stress * Area = Load

260 lb/in. sq * 3.976 in.^{2} = 956 lb

Round to 1000 lb for calculation purposes

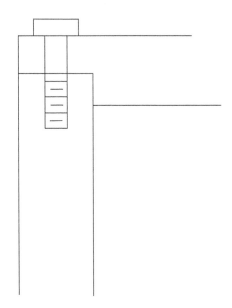

FIGURE 3.2
Cylinder/fastener assembly.

Select a design factor of 5

$5 * 1000 \text{ lb} = 5000 \text{ lb}$

Design four screws equally spaced and loaded around the cylinder. The validity of equal fastener loading should be verified by the designer: four screws, each carrying 250 lb of peak working load. As a design alternative, we could have designed two screws each to carry 500 lb of load, two screws and two pins, or two screws and two machined tenons on the cylinder head casting and two untapped receiving holes in the cylinder body, thereby possibly saving assembly costs of fastening materials and assembly time. A good visual diagram for recording and observing the fastening actions and reactions in the cylinder assembly is with a joint diagram that records the forces applied and the resulting dimensional changes in the assembly.

As a caution to the reader—let it be well understood that this is an idealized analysis. These calculations represent usually safe but simplified rules used by designers to size the joint members and fasteners. They are presented to demonstrate mechanical technique. The actual behavior of the joint is more complex than is suggested here. Critical joints affecting safety or cost may often require a more in-depth study and knowledge.

Figure 3.3 shows the cylinder head cap and body. As the forces increase, the assembly compresses, as noted by the letters F and C. If we make a second diagram for the screw in each corner of the cylinder head, we see one like that in Figure 3.4.

As the screw is put under load, either by tightening, or by application of force by cylinder head pressure, it elongates, or increases in length. I find that even with experienced assemblers, these facts of fastener elongation and component compression are often not understood or considered in operation; yet they play a key role in understanding tension fasteners. It is the action of fasteners elongating, and assembly member's clamp-affected zones (CAZs) compressing that allow these deformations, usually within the

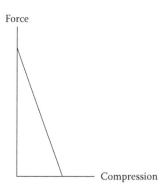

FIGURE 3.3
Force–compression diagram.

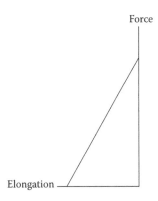

FIGURE 3.4
Force–elongation diagram.

general elastic limits of the fastener and assembly materials, which develops the fastening action.

If we put the two diagrams side by side, we have a good graphics of the assembled components and fasteners, under load and at work, as shown in Figure 3.5.

As an idealized model of this fastening action, we can view the model material deformations when the fastener is preloaded and applying clamp load.

Now let us turn our attention to how we achieve the clamp load in our pneumatic cylinder fastenings. We will first list our site data:

Fastener selected: #10-32 × 2 in. SAE grade 5 steel cap screw, zinc/clear chromate plated

Tapped internal thread: Class 2B internal machine screw with sufficient depth that a 3 diameter thread engagement will not cause a bottoming or interference

Thread friction index: estimated at $K = 0.2$
External thread stress area = 0.02 in.^{2}

First, we will establish how strong the screws and internal threads are when they are under load, engaged, and lengthened/compressed. For the screw, we look up the grade 5 strength level and find 120,000 pounds per square inch, or psi. While this is useful information, it is the proof stress of 85,000 psi that is of more use to us as we plan our assembly work. The proof strength of a tension fastener, as determined by the proof stress of the fastener material and its geometry in our cylinder head fasteners, is that load below which no significant deformation or elastic deformation occurs when the load is removed.

Calculating that load limit for our #10-32 screws, we obtain the following:

$$\text{Proof load} = 0.02 \text{ in.}^2 \times 85{,}000 \text{ lb/in.}^2$$
$$= 1700 \text{ lb}$$

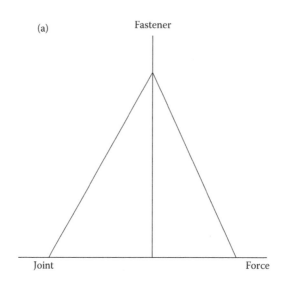

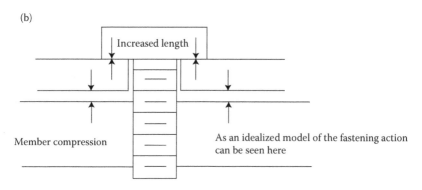

FIGURE 3.5
(a) Joint diagram, elongation, and compression. (b) Tension fastener applying clamp load.

Each of our screws can withstand a 1700-lb tensile load without going into yield and permanently deforming. We next look at the internal threads and by observation see that the area "at work" or the CAZ is by geometry significantly larger than the external screw threads. We have also decided to have a length of engagement between the external and internal threads of 3 diameters, or in other words, three times the thread nominal diameter. In this application that means:

Thread nominal diameter = 0.190 in.
Thread engagement length = 0.57
Thread engagement = 3 diameters

Extremely deep thread engagement beyond three diameters of threads results in diminishing returns in performance. The full-form threads closest to the head bearing area carry proportionately a higher load than those that are farther away. This is a function of the relative ductility of the engaged threads and their spacing or pitch.

As a test, we can make up some test articles and on a tensile tester first load a section of pneumatic cylinder cap, cylinder body, and #10-32 grade 5 screw first to 1700 lb and hold, then on to ultimate tensile yield, and then to fracture. This would establish the engineering validity of our fastening data and also point up the assembly section of least strength. This can be useful in highlighting areas for strengthening during design and manufacturing. In our application, the screw threads would be one such area. In rigid metal joints, this is again considered good practice.

Since we will be using a wrench at the assembly line to install and preload our cylinder head screws, we will calculate the estimated wrench torque needed to preload the assembly and its fasteners. This will help us estimate the turning forces and power required to install our fasteners. It also gives a useful benchmark for the work input required to do this assembly. The torque/tension equation we will use is as follows:

$$\text{Torque} = \text{Friction} \times \text{Thread diameter} \times \text{Clamp load}$$
$$T = K \times D \times P$$

with
   T in inch-pounds
   K (dimensionless)
   D in inches
   P in pounds

Since we anticipate that our build rate for these pneumatic cylinders will approach a relatively high rate and that statistically we expect some variability in the engineering value of the fasteners, components, lubricants, and assembly conditions, we will plan to tighten to a percentage of 75% of the proof strength load of 1700 lb. Our value for P will be set as follows:

$$P = 0.75 \times 1700 \text{ lb}$$

$$P = 1275 \text{ lb}$$

Completing our calculation of tightening torque, we compute:

$$T = 0.2 \times 0.190 \text{ in.} \times 1275 \text{ lb}$$

$$T = 48.45 \text{ in. pounds}$$

Converting and rounding, we can also say 4 foot pounds.

A tightening torque of 4 foot pounds, or 48 in. pounds, will develop and apply approximately 1275 lb clamping force on our cylinders. If this estimated clamp load is not of sufficient accuracy, we can perform a series of torque tension tests and also calculate and plot a possible range of clamp load performances. Equipment used would be torque wrenches and tensile testers, torque and tension transducers, or similar apparatus.

Several useful pieces of equipment are available on the market to allow us to determine the actual clamp load developed in test specimens for a given tightening torque. One method that would be useful if special torque/tension transducers or load cells were not available would be to use a standard tensile tester. As mentioned above, on occasion the author has tested the screw, nut member, and a section of the assembly for the torque input–tensile clamp load output, which was reported to the customer, and to determine which was the weakest section of the fastening system for further development work. On one application, an engineering client wished to determine the frictional coefficients of three different fastener finishes on their high-strength screws. Again, a tensile tester and a set of tapped fixtures were used. Figure 3.6 shows a typical setup for torque–tension fastener plotting.

It is instructive to consider the variability of clamp load to changes in the friction coefficient K, which is composed of friction between the internal and external threads and under the head of the screw in the area

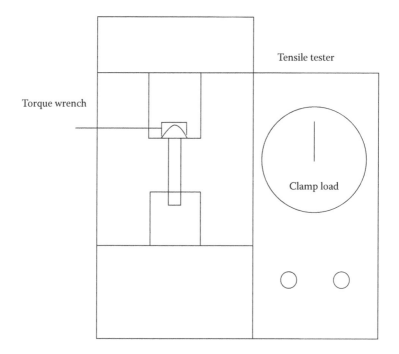

FIGURE 3.6
Tensile tester/torque wrench plotting setup.

known as the bearing area. Let us look at the clamp load developed in this same application with five different K values, all of which can be found in everyday industrial fastener assembly applications. Note that we will apply the same level of preload in each case, 1275 lb, but with the required torque to achieve this preload, increasing as the K value increases in each application.

1. $K = 0.3$
   $T = K \times D \times P$
   $\quad = 0.3(0.190 \text{ in.})(1275 \text{ lb})$
   $\quad = 72.7 \text{ in. lb}$

2. $K = 0.25$
   $T = 0.25(0.190 \text{ in.})(1275 \text{ lb})$
   $\quad = 60.6 \text{ in.lb}$

3. $K = 0.2$
   $T = 48.5 \text{ in. lb, from previous calculation}$

4. $K = 0.15$
   $T = 0.15(0.190 \text{ in.})(1275 \text{ lb})$
   $\quad = 36.3 \text{ in. lb}$

5. $K = 0.1$
   $T = 0.1(0.190 \text{ in.})(1275 \text{ lb})$
   $\quad = 24.2 \text{ in. lb}$

So it can be seen that a change in tightening friction from 0.3 to 0.1 would mean a reduction in torque required to achieve our target preload. A value of 0.1 can be typical with very efficient extreme pressure-type lubricants. A $K = 0.3$ or higher would not be uncommon for gritty threads. So, it is important to know the coefficients of friction for the threaded fasteners in your assembly; it will as well be good practice to control elements such as more-efficient-than-planned lubricants, or their opposite, friction-increasing abrasives, or grit in the fasteners and threads. Of particular use may be a torque-clamp load plot to serve as an operating plan for the assembly. If we plot all five of the K values we have just calculated for our pneumatic cylinder screws, we would see a chart similar to Figure 3.7.

As an observation, it is possible in some softer metal assemblies' joints that the torque-clamp load relationship can be nonlinear due to distortion of the mating threads as thread load increases with tightening torque. In these cases, it is obviously important to know and quantify these values to the extent possible. Again, torque–tension testing under real or near-real application conditions can prove very beneficial in these applications.

Once the target preload is selected, and we have indicated this to be 1275 lb in our idealized example, it will serve us well for illustration purposes to look

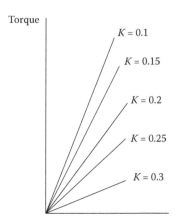

FIGURE 3.7
Torque-clamp load plots.

at this preload as a vector quantity. We will first draw our pneumatic cylinder screw clamp load as a tensile vector. We can consider this a "preload." It is the "rubber band" elastic stretch to hold the deck of cards together. Next, we will compare the service load from the operation of the cylinder by the buyer and user of the cylinder and offer some technical observations.

Our service load vector can be simply represented as illustrated in Figure 3.8.

Then we can represent the cylinder's fastener preload on the screw in a similar manner, as in Figure 3.9.

By comparing the two forces, and the resultant stresses and corresponding strains, we can determine the integrity of our fastener system. This comparison approach simplifies the comparison of forces which are coaxial to the screw and tapped hole. In actual practice, it is important that any

500 lb

FIGURE 3.8
A 500-lb tensile service load.

1275 lb

FIGURE 3.9
A 1275-lb service load.

secondary forces be calculated into the fastener force system and that no shear or bending forces of significant magnitude be present, which reduce the fastening strength. In our case, the two forces can be compared as in Figure 3.10.

In fact, the behavior of the joint is not this straightforward. Where safety is an issue, more rigorous joint stress analysis may be advisable.

To reinforce the concept of preload and its value, let us consider our assembled cylinder in use. At startup, the fastened site has a fastened load of something approaching 1275 lb. During operation, residual preload loss due to material relaxation or a phenomenon known as "bedding in" can diminish this preload. On starting our cylinder and performing work with it, the tensile force exerted on each screw is 250 lb, so long as we assured equal loading to the extent practical during design and assembly. We could have taken a good step in this direction by tightening the four screws in a crossing pattern using several increments of increasing tightening torque until we reached our 4 ft lb preload torque.

With 1275 lb of preload already loading each screw site, it is unlikely the 250 lb service load will unload the fastener. Rather, it opposes a percentage of the preload, in this case 250/1275, or slightly less than 20%. If we consider the goal of holding the pneumatic cylinder head clamped firmly to the

FIGURE 3.10
Overlay of service and preloads.

cylinder, under the current conditions, and continuing our simplified, safe analysis, it would take another 1025 lb of force on each bolt before the cylinder head started to separate from the cylinder.

Another way to evaluate the preload versus service load application in rigid metal joints is to consider strain, or fastener elongation, rather than force since they are both results of the same loading. At 1275 lb of preload from our tightening torque of four foot pounds, we have an elongation in the screws, which can be approximated by the equation:

$$e = F \times L/A \times E$$

where
  e is the fastener elongation
  F is the fastener force
  L is the fastener grip length
  A is the fastener stress area
  E is the fastener modulus of elasticity

For our application, the equation would calculate an elongation:

$$e = 1275 \text{ lb } (1.250 \text{ in.})/0.02 \text{ in. } (3.0 \times 10^7 \text{ psi})$$
$$e = 0.0027 \text{ in.}$$

Our cylinder head screws elongate 0.0027 in. when tightened to four foot pounds for the conditions we have established. Like a spring, our screw is stretched elastically and clamps its gripped assembly. If we look at the elongation from a 250-lb service load with no preload, the fastener elongation would be as follows:

$$e = 250 \text{ lbs } (1.250 \text{ in.})/0.02 \text{ in. } (3.0 \times 10^7 \text{ psi})$$
$$e = 0.0005 \text{ in.}$$

The elongation or strain from the service load alone is less than from the preload. Recalling again the stress–strain diagram for steel shown in Figure 3.11, in this case of our pneumatic cylinder screws, with negligible added strain from the service load, the fastener should experience small and absorbable amounts of service stress.

As a goal in rigid joints, the fastener preload is greater than the service load. In rigid tension joints such as our hypothetical, idealized example, preloading above the service load is good practice toward providing protection against joint separation. By preloading above the service load, the fasteners are in a position to provide a well-fastened assembly.

It is important to note here that in addition to tension joints in rigid metal application, shear joints are used. The most common applications are in

Stress, psi or MPa

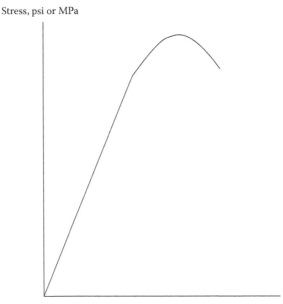

Strain in./in. or mm/mm

FIGURE 3.11
Stress–strain diagram.

airframes and in steel construction, where the friction from plate surfaces clamped together provides the holding force. In many industrial assemblies, shear features such as pins or shouldered features are designed into the fasteners and/or assembly to carry the shear loads, while the tension fasteners provide the clamping force.

In considering elongation, allowance for bedding in or other types of time-functioned joint-holding force relaxation should be provided for in the fastener design. In the previous example, at preload a 0.0027 in. elongation was provided. If the assembled cylinder head and cylinder joint faces had some roughness from machining, the wearing down of these high spots would result in a slight decrease in the joint thickness and allow the cylinder head screws to release some of their elongation and also lose some of their preload.

A good way to think about this is as a percentage. If bedding in of 0.0009 in. occurred, 33% of the preload has been lost. Since bedding in or initial wear can be significantly higher in some production volume assemblies, it is beneficial to many assemblies to provide for as much elongation as practical. As an example of techniques that can be employed, a look at the elongation equation will provide us with guidance:

$$e = F \times L/A \times E$$

For a given preload (F), an increase in grip length will increase tension fastener elongation. For this reason, extremely short screws should be used judiciously. Similarly, reducing the grip area of a tension fastener with a reduced shank area, perhaps at or below pitch diameter, can provide elongation benefit provided it is not serving a locating or shear function. As can be seen, since it is an inverse function of elongation, a grip area decrease gives an elongation increase. For this reason, short, large-diameter screws are to be especially installed with an equal measure of caution. If the number of joint surfaces and their surface roughness indicate a bedding-in rate higher than normal, longer fasteners in tensile preload will provide a higher preload retention rate.

Let us consider the functioning of the engaged external and internal threads under preload and service load in the assembly. Machine screw threads, whether inch or metric, are 60° included angle triangular elements with major, pitch, and minor diameters. The threads have a pitch, which is the distance from a point on one thread, such as the thread crest, to the same point on the adjacent thread.

The threads are offset to the horizontal plane normal to the axis of the shank by the helix angle. For many machine screw threads, the helix angle is in the 4° plus some number of minutes range. Coarse threads, those with a larger pitch distance, have a correspondingly large helix angle compared with finer pitch machine screw threads. The pitch is also the distance the fastener moves for each 360° rotation, unless constrainment causes permanent thread deformation, or plastic displacement. If the threads have multiple pitches, having two or more thread starts, the pitch is the same for each thread, but they can be thought of as in series, with each thread building on the advance per rotation of the others.

The two surfaces of the thread with a triangular cross-section function differently under tensile load. Most unified inch series and metric threads have a 60° included angle. The surface facing up to the head bearing area is the pressure flank under typical tensile loading. The flank facing the fastener point is the nonpressure flank.

Let us look at screw threads engaged and under load, as in Figure 3.12.

The load applied by the fastener is carried in the threads on the pressure flank surfaces. They are the areas on the internal and external threads, which transmit the load into the minor diameter and root area of each of the threaded parts. There is some deflection of the threads. This deflection is a function of the ductility of the thread material(s), the thread fit, and the pitch of the thread. In general, it is thought that fine pitch threads deflect less than coarse pitch threads, all other things being equal. Under load, the first full set of engaged threads carry a high percentage of the applied preload, along with an equally high percentage of any service load. As we have established, in relatively rigid joints our goal is to have the preload override any service load. For a relatively ductile set of fastener/assembly joint rate materials, a thread load distribution for coarse and fine threads might look similar to the following description.

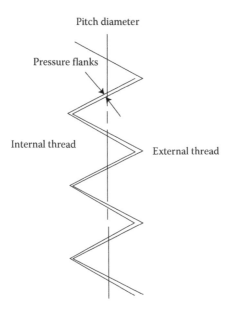

FIGURE 3.12
Thread fit.

Here again, where the application warrants, in-depth stress analysis may prove a beneficial fastening engineering step.

	Load Percentage Carried by Threads	
	Coarse Thread	**Fine Thread**
Thread 1	35	29
2	25	21
3	17	15
4	12	11
5	11	8
6		6
7		5
8		5

As can be seen, the number 1 thread, which is the full form, engaged thread closest to the screw head and threaded hole entrance, carries the highest percentage of the clamping load in each application. The fine thread has a more uniform distribution. It is for this reason that it is normally a case of diminishing returns to engage more than three times the thread basic diameters (or 3 diameters as it is sometimes called). The engaged threads furthest away from the more highly stressed upper threads carry a negligible part of the clamping load. If any thread pitch, flank, or helix error is present, the opportunity for thread scuffing is more likely present. And preloading requires more fastener turns and time.

As a rule, 2–2.5 diameters engaged for higher spring rate materials and 3 diameters for more ductile materials is normally good practice. Before we leave machine screw thread engagement, let us look at diametral engaged thread fitup. In the inch system threads established by Canada, the United Kingdom, and the United States, and laid out in ANSI B1.1, there are three primary thread classes for both internal and external threads. Setting them side by side, they are

	Internal	External
Loose fit	1B	1A
Medium fit	2B	2A
Tight fit	3B	3A

There are nine possible, free assembling class combinations that can be chosen by the assembly designer. A 2A/2B or 3A/2B fit up combination would be normal for most assembly work. If plating is involved with the external threads, a 2A before plating/3A after plating is normal practice. For precision work, 3A/3B is close to line-to-line fit. At tension, the pressure flanks of the engaged threads increase in pressure reaction to the clamp load, while the nonpressure flanks may have clearance. An interesting study the author had the opportunity to review some years back tested the vibration resistance of several sets of threaded fastener assemblies, the only variable being diametral thread fit. The report found no significant difference in vibration resistance for medium or tight fitting thread combination with the conclusion being that thread pressure flank load level was the controlling factor of vibration resistance.

Also available are intentional interference fits, which greatly increase the friction in the threads and thereby provide vibrational loosening resistance. These torques are recorded as prevailing-on for tightening and prevailing-off for loosening with some loss of prevailing-off torque with the number of uses. Good data for the locking fastener type you are interested in for your application would be prevailing-on, prevailing-off first use, and prevailing-off fifth use torques to show the loss of vibration-loosening resistance ability with usage. A good locking thread system would provide relatively low prevailing-on torque, relatively high prevailing-off torque, with the fifth prevailing-off torque showing good retention. Well-lubricated steel threads that subsequently rust with time and use provide this. Of more practical use are thread-locking compounds which cure anaerobically.

Metric thread diametral clearance involves the same geometry of 60° machine screw threads with major, pitch, and minor diameters. The thread callouts and thread classes offer more information and allow for plating in the standard classes.

Internal thread classes are noted with a capital letter, H, G, and so on. While external threads are noted with lower case letters such as h or g. This is in accordance with Systems International (SI) practice for limits and fits. Only two classes of metric threads are commonly used.

	External Threads	Internal Threads
Medium fit	6g	6H
Tight fit	4g6g	4H5H

For plated threads, classes 6g and 4g6g provide an allowance that is an improvement on the inch system tolerances. ANSI B1.13M has formulas for computing the dimensions and limits of metric threads.

An important point to note in the production of internal threads by tapping is the function the tapping tool plays in the final class of internal thread produced. In additional to being supplied in the three geometries of taper, plug, and bottoming to perform the tasks of starting, tapping, and finishing, taps are supplied in diametral limits. A typical limit would be #10-24 GH3. The G signifies that the tap was ground. This is somewhat redundant since most quality taps are now produced with this method as opposed to turning. The significant part is the H3. This signifies that the tap is designed to generate a #10-24 internal thread, which is 3 units "high," or above the basic diameter limits for an ANSI #10-24 thread. Each unit is 0.0005 in., so a GH3 is 3×0.0005 in. or 0.0015 over basic when new. Of course as it wears in production, the tap holes will become increasingly smaller. Care in the selection and maintenance of threading tools will allow you to design and manufacture threads to the tolerances your application requires.

On fasteners with geometries other than machine screw threads, the goal is the same. A joint fit should be specified and maintained throughout production. The goal is high joint efficiency, joints as strong as or stronger than the component materials they are assembling.

3.2 Nonmetal Joints

In some ways, nonmetal assembly components offer a degree of flexibility that the metal joint does not. There are not as many time-proven and developed rules to guide our specifications. There is also an established base of suppliers of fasteners for nonmetallic components applications. Let us specify some of these fastener types for a hypothetical product to allow us to review some of the mechanics involved.

First, let us consider a panel of a composite material that has been formed into a contoured structural member of a larger end product. Our goal will be to fasten several injection-molded plastic brackets to the panel.

3.2.1 Snap Fit Assembly

Our first joining technique will be to manufacture the panel and brackets for snap fit assembly. The factors which we will need to consider will be the geometry of our snaps and receivers, their tolerances, and the resulting fastening load retention. For this case we will design a rectangular snap with a barbed end and rectangular receiver in our panel. Figure 3.13 shows a design.

Some of our design concerns will be the shape and angle of entry of the snap fit ramps and the force required to snap the fastener together. Moreover, if removal is required, we may want to modify the snap so that only one barb is in engagement, with some clearance opposite the barb in the snap zone for releasing the barb. The tolerance of the molded snap geometries will be a factor in that they will influence fastening load retention. Also, we want to determine if any service load conditions such as service temperature will decrease the snap fastening's strength due to elevated temperature's softening of the plastic, lowered temperatures making it brittle, or ultraviolet radiation from the sun causing plastic degradation.

3.2.2 Adhesive Assembly

If we decide to use an adhesive fastening approach we want to consider the surface area and its condition as well as how we apply the adhesive. Let us consider the fastening site in Figure 3.14.

Our concerns here will be the surface preparation of the mating surfaces, the thickness and condition of our adhesive, and the compatibility of our adhesive for its intended service. We can test the cured joint by applying force to the joined surfaces. We can test the pull strength with a force and

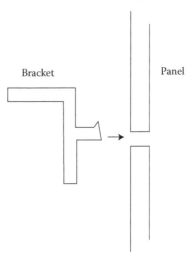

FIGURE 3.13
Snap fit assembly sketch.

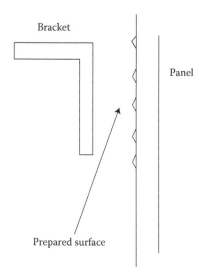

FIGURE 3.14
Adhesive assembly sketch.

reaction normal to the adhesive-held surfaces. We can test the adhesive shear strength with forces parallel to the joint. A loading of a combined set of these stresses would give us a good profile of our adhesive joint's "cured" strength. To test for time and temperature effects on joint strength, we could vary the joint temperature as well as set aside prepared specimens for testing during and after time and temperature.

3.2.3 Blind Rivet Assembly

Now let us consider the mechanics of a blind rivet attachment of our assembly, as shown in Figure 3.15.

Some of the mechanical considerations will be the compressive strengths of the bracket and panel material. We will also be interested in the force required to first cinch up the joint and then break off the mandrel. If the latter forces exceed the compressive strength of our joint materials and their clamped areas, the mandrel may pull through the rivet body rather than breaking off cleanly for a clean rivet installation. A remaining consideration will be rivet clamp load retention. As has been emphasized previously, test riveting under a range of conditions and the recording of results and observations can serve the assembly well in addressing these concerns.

3.2.4 Self-Threading Screw Assembly

To round out our assembly approach, let us consider a self-tapping screw to secure our panel and bracket. Figure 3.16 shows a sketch.

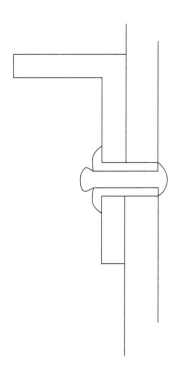

FIGURE 3.15
Blind rivet assembly sketch.

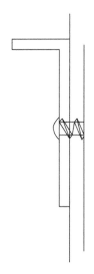

FIGURE 3.16
Self-threading screw assembly sketch.

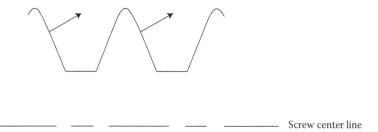

————— —— ————— —— ————— Screw center line

FIGURE 3.17
Screw thread forces. Increased length 0.0034/in. of grip length per 100 KSI prestress with steel
fastening materials.

Our mechanical consideration with a self-tapping screw installation will
be with the strip-to-drive ratio of the screw and fastening site. We will first
make sure that the size of the screw's thread and body diameter fit well into
the hole site, both the through hole in the bracket and the anchoring hole in
the panel. We also want to make sure that the thread spacing of the screw
thread as well as the thread geometry is suited for the panel material.

A view of the thread forces exerted by the thread-forming thread on the
panel is shown as in Figure 3.17.

The screw thread captures panel material in the area between the major
diameter and root diameter. The applied thread load is relatively perpendicular to the thread half angle. For this reason, thread-forming threads for
plastic can often benefit from thread angles narrower than the machine screw
of type A or B tapping screw thread angle of 60°. A thread angle between
30° and 48° can often give good performance in relatively soft fastening site
materials.

A good test is to prepare some holes in coupons of the proposed site material. If the test holes are made differently from the production holes, say
drilled for testing and molded for production, factor this into your data. A
difference of 15% higher drive torque for molded versus drilled holes has
been observed. Record the torques needed to drive the screws and form
threads in your application. Then continue applying torque until stripout
occurs. Make a ratio of torques. This is the strip-to-drive ratio. A ratio of 3:1
is considered good practice.

3.2.5 Some Additional Thoughts

In planning any fastened assembly, two points can be of great help in achieving manufacturing joints that are efficient, joints whose strength approaches
or exceeds the strength of the component materials, and joints that require
only reasonable amounts of fastener and installation costs.

First, consider the engineering mechanics of how the candidate fastening
will apply clamping forces as well as where and how the reactions will be

carried. And then test with as many application-specific conditions and with as much laboratory rigor as the integrity of the final assembly requires.

3.3 Applications View of Fastener Functions

Following up on the idea of testing fastener assemblies, both the laboratory and the field provide useful applications, which we can review to increase our understanding of fastening mechanics. A useful concept to introduce is the idea of CAZs in fastened applications. The relative spring rates, or springiness of both fastener and fastened assemblies members is introduced in fastener mechanics diagrams. A typical diagram can be seen in Figure 3.18. A cap screw and pump members are being fastened. The force required to elongate the bolt grip to design preload is drawn on the vertical axis. The elongation of the bolt grip is drawn on the left horizontal axis. The compression of the pump CAZ is drawn to the left. In the analysis of threaded fastener joints, this comparison of the spring rates of fastener and fastened is a fundamental

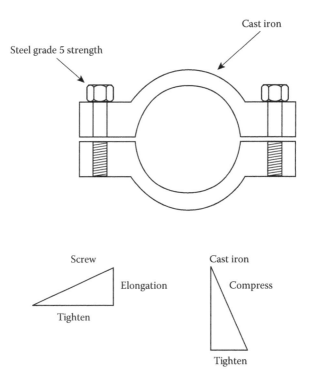

FIGURE 3.18
Elongation and compression.

step toward understanding the mechanical operation of threaded assembles. It is important to note that not all assemblies have screw thread fastening, as we saw in the previous chapter with riveted, glued, and welded assemblies. That said, fastening with screw threads remains a dominant force in industrial production. In expanding our knowledge of threaded fastening mechanics, let us use the relative spring rates of the components as depicted in Figure 3.18, with a reading or further analysis of the threaded connections using the concept of its natural CAZ. Note that in Figure 3.18, the slope of the assembly member and fastener length deltas, the slope represents the springiness of these components. We can make a ration of their slopes to provide a composite index. A ductile material, such as low-carbon nonheat-treated steel is very springy. This type of fastener material would be found in common machine screws. Cast iron, such as an internal combustion engine block, is not very ductile. Its slope would be much steeped, with less compression under screw clamp preload. For a first example, let us use a well pump. The one in my home is a jet pump driven by a 3/4 horsepower electric motor. For discussion, let us say it is clamped to the cast iron pump housing with four hex cap screws threaded into the pump body. From the head markings, we can see three radial hash marks. These let us know that the cap screws are grade 5 strength level. Adding two to the number of hash marks on the head gives us the screw's strength grade. Figure 3.19 shows a sketch of our pump fastening site. Note that the screw has a mass as does the surrounding pump body. From the grade 5 screw material, we know that the screw is manufactured from medium-carbon steel with a ductility of at least 14% elongation and a reduction of area of at least 35%. We know that our pump housing material will have a much lower ductility. Let us establish it at 5% elongation and 18% reduction in area. This would be indicated by a slope for the screw at a much lower angle that the CAZ material of the pump. In Figure 3.20, a possible CAZ for this assembly site is mocked up. These two concepts, used together, can offer a useful model for better understanding the mechanics and fastened assemblies. The relative spring rates of fastener and fastened component's materials provide a view of the material performance

Consider the spring rate of the assembly

FIGURE 3.19
Step 1.

Consider the spring rate of the fastener

FIGURE 3.20
Step 2.

capabilities. A view of the CAZ of these materials under both installation preloads and service loads over an operating range provides a view of these materials in use. Material spring rates can be obtained from tables of material properties. The components themselves can also be tested in the materials laboratory. There are a range of methods that can be used to estimate the CAZ areas. Especially on larger assemblies, strain gages can be used to measure the stresses in individual location in the CAZ. Similarly, finite-element analysis modeling software can divide the assembly into finite nodes. Each of these nodes can then be analyzed for the fastening stresses and resultant strains in its node. Of importance from an application engineering point of view is the ability to observe the components with an eye for how the CAZ might be operating under expected and extreme service conditions. In Figure 3.21, we sketched three of the many CAZ shapes that might occur. The barrel shape has the maximum area located mid-zone, with the stresses building and then subsiding in either linear direction, coaxial to the screw centerline. The rate of stress change is exponential. In the conical shape, we find a similar stress distribution in the CAZ. The rates of clamping stress increase and decrease are more linear. In the third CAZ shape, the clamping stresses build rapidly and are distributed at rapidly diminishing magnitude with distance

Evaluate them both as an system

FIGURE 3.21
Step 3.

from the bearing surface of the fastener head. As the reader can intuit, a wide range of CAZ shapes is possible. This third shape is one I have observed in numerous fastener applications. The starting internal threads are less than full form. The pressure flank contact areas do not reach full contact area until a short distance into the internal thread. At this location, a large percentage of the clamping load results in stresses developing in the clamped part which directed up at a 30 degree angle and out at a 90 degree angle into the CAZ, provided sufficient material is present in the site area and of sufficient strength to provide a suitable reaction. The stresses then decline in magnitude. For our next example, let us look at first a bolt, clamped plates and nut, then add a washer under the nut, then finally a washer, nut, and jam nut. These are shown in Figure 3.22. The first application will give us a CAZ with a shape similar to the rapid build, slow decline shape discussed above. Our hex nut will have a lead-in thread section. The first, full-formed thread engaged will be the clamp zone location of highest stress. As the outside is hexagonal, the corners and flats also will provide radial stress gradient pattern in the horizontal plane. Next, we add a flat washer. This will increase the area brought into the CAZ. With the area increased for a given clamp load, the peak stress will be reduced. The effectiveness of this generally desirable effect would need to be weighed against the increase in parts count. In our third example, the addition of a jam nut, torqued to a percentage of the torque of the full nut, elongated the shape of our clamp effected zone as well as distributing the stresses over more thread flank area. It is noteworthy that one internally threaded nut with the same length of engagement would not provide the same benefit with respect to clamping stresses on the thread pressure flanks and CAZ shape. Once again, parts count versus CAZ shape would need to be evaluated for economic economics.

A very successful race car builder was known for being able to make these fastener economics versus performance decisions with remarkable precision. The goal being to have a race car that was capable of operating under race stresses for just the length of time to complete the 500 miles or whatever length the race was to endure. In those applications, the additional mass in

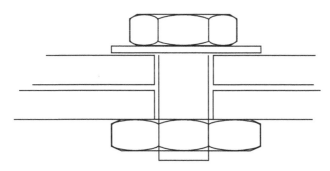

FIGURE 3.22
Bolted assembly.

assembly components to provide even one additional lap of fastened assembly endurance was unwanted weight. In most fastened applications, more assembly integrity will usually be more than justified.

Let us consider a 12-point flange nut and 12-point bolt. This assembly would be found in aircraft and other high-performance, high-integrity assemblies. The material properties of these components will be similar in metallurgy. The joined components are often dissimilar. In an application that the author has observed, a titanium 12-point bolt is installed in a flight-critical aircraft assembly. It joins two metallic plates clamping on either side of a carbon fiber member. The 12-point nut is also of titanium, with an all-metal locking thread. The washer faces of the bolt and nut define the end limits of the CAZ. The relative material properties of the carbon fiber member set the parameters for the CAZ geometry. A study of the performance of this assembly could include a plot of the spring rates of the bolt, nut, plates, and carbon fiber member as well as a plot of expected and possible clamp-effected zone. These fastener engineering tools can provide a good foundation for understanding the behavior of this or any assembly. Using fastener engineering tools with a graphic output provide a readily communicated visual representation of the assembly. A sketch of this assembly with an estimated CAZ is shown in Figure 3.23. One possible use of this CAZ analysis would be for the design and manufacturing of the carbon fiber clamped member. Its fiber direction and winding pattern can be oriented with respect to the optimal direction and pattern for clamp and service load-carrying performance.

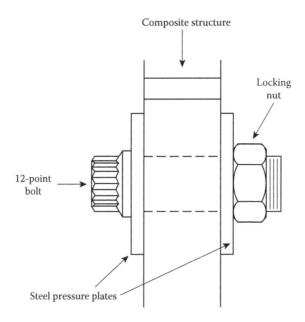

FIGURE 3.23
12 Point bolt assembly with dissimilar materials.

3.4 Fastener Strength of Materials

The mechanics of materials provides a scientific approach to the understanding of materials for fastening, joining, and assembly work. Material strength is a fundamental factor in the performance of assembly components. The equation $S = P/A$, or stress equals load divided by area indicates that for a given load and dimensional size of a component, a proportional stress occurs in a fastening. For the fastening to do its intended job, its material must have sufficient strength to carry this load without failure.

Taking an overview of the study of material mechanics, one method to approach the subject is to first look at a material on a micro level in a free state, then at normal size, and then under load. Microscopically, say with a scanning electron microscope, we would see the grain surfaces of a metallic fastener material. With alloy steel 4037, we would have surfaces in contact composed of iron, carbon, manganese, and molybdenum, as well as other trace elements. What holds these in molecules together? The negative charges of the electron shells and the positive charges of the protons in the nucleus provide the tensile field that makes splitting these individual units of matter require more energy than is normally available in routine collisions. Consider the high-strength material, Inconel 718. Nickel, iron, and other metallic and nonmetallic element's individual matter of atoms and molecules are bound together with electronic charges strong enough to provide much usable strength. When processed into ingots and billets, grains of Inconel 718 of more complex geometry form the larger structure, which can be used for fastener and component manufacturing. It is at the surfaces of adjoining grain boundaries that strength-critical "weak links" are typically located. With the aid of powerful magnification, we could see the grain structure of the material. Using very small units called angstroms, we could classify fastener and other component materials as either coarse or fine-grained. The coarser the grain, the longer the grain boundaries. Most metallic materials have a grain size that yields more strength than others. Cold working and heat treating, as well as in-service forces, can alter grain boundaries. The inner molecular and grain boundary strengths of a material such as Inconel 718, or molybdenum alloy steel such as 4037, can be thought of as the "micro" strength properties of a component. A widely used fastener material such as type 4037 alloy steel contains nominally 0.37% carbon, as well as manganese, molybdenum, and other "trace" elements to give it its micro-properties (see Chapter 7, AISO 4037). The grain structure of this material can come in several microstructures. Three typical ones are shown in Figure 3.24. The spheroidal structure is typical of raw material used in cold heading of fasteners and components. The grains have rounded boundaries, which increase the formability of the material. Let us look at a stress–strain diagram for a typical low-carbon steel, type 1022, in Figure 3.25.

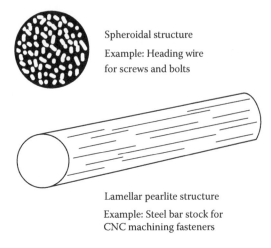

Spheroidal structure
Example: Heading wire
for screws and bolts

Lamellar pearlite structure
Example: Steel bar stock for
CNC machining fasteners

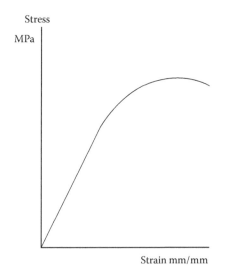

Fine tempered martensite structure
Example: Heat-treated fasteners

FIGURE 3.24
Grain Structure.

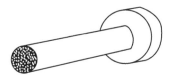

FIGURE 3.25
Stress–strain diagram.

Notice that the strain, or change in shape from increasing levels of stress, is a straight line up to the yield point. If we increase the stress, the change in form, or deformation, increases. If we remove the force and decrease the stress to zero, the shape of the material returns to its original shape. In actual practice, there are always residual forces and stresses, from gravity and temperature, as well as others. Consequently, there is typically, local yielding of most materials at their more highly stressed areas. Beyond a material's yield point, the stress to cause a given amount of deformation is lower; however, the shape change is more permanent. Removing the stresses does not return the material to its preforce shape. It is a kind of an end to innocence for the material. In classic strength-of-material terms, deformations below the yield point are elastic. Beyond the yield point, they are plastic. Elastic deformations snap back. Plastic deformations flow into new shapes.

In cold heading, forming, and stamping, the manufacturing process operates in the material region between yield and ultimate tensile as seen in the stress–strain diagram. And the spheroidal structure gives the manufacturer more room in the region to form. As an additional note, with working of the material, this zone between yield and tensile "work hardens" and the forming zone narrows. Beyond any material's ultimate tensile strength, cracking or material failure occurs. Consider the side-by-side diagrams of the 1022 and Inconel 718 materials before and during forging in the stress–strain diagrams in Figure 3.26. For this reason, fastener and component geometry, and forming process forces and speeds are critical to avoid cracking and discontinuities, which can lead to service failures, especially in higher-strength materials. In lamellar metal microstructures, the grains are oriented more like the layers seen in a sheet of plywood. This can be the best structure for machining processes where the separation of a smooth chip results in an improved machining index and can provide increased cutting tool life. Finally, the martensitic grain structure, which is not ideal for either forming or machining, is often

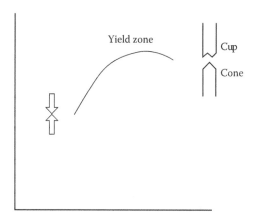

FIGURE 3.26
Stress–strain diagram with test piece.

optimal for carrying load. This grain structure, with a minimum of retained amount of another grain structure, austenite, is the desired result of heat treatment of medium carbon and alloy steel components. Referring to the stress–strain diagram again, we can define several mechanical properties. In Figure 3.26, we can see the start of yield, the ultimate tensile strength and the breaking strength. This last is lower since the material starts rapidly reducing in area, known as "necking down". The proof strength area is identified, so named since the test piece must prove it can carry this load to qualify. The linear rate between zero stress and yield is known in engineering terms as the modulus of elasticity. The modulus of elasticity is expressed in inch units of pounds per square inch, or psi. Hardness is resistance to penetration, usually with some indenter such as the Rockwell "C" indenter or the Rockwell "B" hardened ball. It is a convenient proxy for tensile strength as hardness is approximately proportional. Hardness cannot as accurately predict fastener dynamic strength. The area under the stress–strain diagram can be an indicator of these dynamic strengths, as listed as follows, and can be thought of as the ability to absorb energy, especially in the CAZ of the assembly.

Impact strength: This is the ability to absorb suddenly applied forces, as defined as in time shorter than a percentage of the component's natural frequency. An example would be a weight dropped from some height onto an assembly. The Charpy and Izod impact strength scales use units of pound-feet.

Fatigue strength: The ability to withstand frequently applied forces without failure for some long time period. The classic time period is 10 million cycles. A great example of metal fatigue is the breaking of a paper clip (itself a fastener) by bending it back and forth (reverse bending). A tension–tension fatigue strength for a good-quality alloy steel screw might be 10,000 lb of alternating load for 10 million cycles.

Discontinuities: These small microcracks, scratches, and pits are the features that can lower fatigue strength.

Cracks: These anomalies can be of a more catastrophic nature. Liquid penetrant and magnetic particle inspection processes, depending on material type, are used by trained operators to detect these defects.

Laps: Folds in the material, especially in the screw threads from thread rolling, which can be stress risers.

Cold shuts: Similar to laps, but also found on other formed areas of a component from uneven or interrupted flow of raw material.

Thermal expansion: This is a fact of all raw materials, according to the material's coefficient of linear expansion. It is especially important when materials of dissimilar rates of thermal expansion are used in applications with significant temperature gradients.

Galvanic couples: Referring to our previous discussion of molecular electronic charges, materials with galvanic electrical potentials of sufficient charge gradient can corrode rapidly if care in design and assembly is not exercised. A classic example is the aluminum automobile panel, gripping a piece a stainless steel trim fastened with a zinc-plated steel screw. Add water and road salt and the chemical reaction of steel and aluminum corrosion is as efficient as if a storage battery was flowing current across the corroding fastening surfaces.

Nonmetallic material strengths have different units than metallic ones. For many plastic components, the strength scale is the flexural modulus, which gives a measure of the deflection of a test beam of the nonmetallic material under transverse load. Another measure is the durometer reading of rubber-like materials. One very common fastener design is the sealing screw that is manufactured with a steel or stainless-steel screw having an inner curved radius in the under head bearing surface. Into this ring is placed an O-ring of Buna N or similar material. The rubber-like O-ring compresses when the screw is tightened into the assembly, affecting a positive seal.

3.4.1 Engineering Model Method

In many engineering applications, including mechanical fastening, the construction of a model of both the component and the service loads helps to better understand the application and improve its performance. Consider the outdoor grill propane tank valve in Figure 3.27.

The model we have drawn shows both the macro-view of the assembly and the gas, mechanical, and ambient forces. The micro-view of the screw threads shows the forces on this critical section of the assembly. The ability to construct and quantify accurate models of fastened assemblies is often a key tool in assembly application engineering work.

As a fastener application student, being able to "read" the force system of a fastened application is a useful talent to develop. You can notice the bridge support bearings of underpasses as you drive in a car. Consider the load from the passing cars over the bearing, the weight of the highway, and other imposed loads such as wind or snow, which are static forces, from gravity,

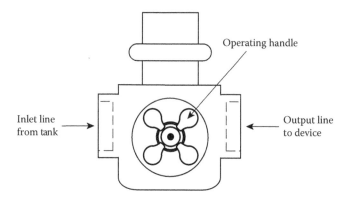

FIGURE 3.27
Propane tank gas valve.

and dynamic forces, such as an overloaded tractor trailer traveling over the posted speed limit. How are the forces supported? What would a plot of the total dynamic forces look like when examined as a graph? How is the energy absorbed? To where does it dissipate? What components in the entire system surrounding the bearing, including its methods of fastening to the bridge and bridge supports, dampening any cyclic forces to minimize vibrations? If a sudden and large load impacted the overpass, say a vehicle impacting it, how would it take the blow? These exercises will strengthen your analytical abilities. They can be practiced almost anywhere observing many types of fasteners and assemblies. They will help you become a better technical person. And this can be fun!

Here are some examples from my experience of fastener applications that provide, hopefully, some ideas for fastener applications that you may encounter in your future work. While formal education in the classroom is critically important, knowledge again in the field augments the theory nicely with real-world experience that any good practitioner needs. These tricks of the trade, or what has been called tribal knowledge, often are first passed on away from school. It is hoped some of these clever parts spark some ideas for uniquely efficient fastener solutions of your own.

Dog point pipe plug: An oil field equipment company required a tapered plug with pipe threads to locate into a hole at the base of the mating internal thread. Undeterred by the fact that no standard existed, headed pipe plug blanks were checked in a lathe and an axial through hole was drilled from the point into the socket. Tenons with a stepped down diameter to provide a slip fit into these holes were also machined. The plugs were threaded and then the tenons press fitted and brazed into the face of the pipe plug. For the small run quantity involved, the cumbersome steps to manufacture these were not an issue. The customer gladly paid the cost to get the special fastener their application required and the job was repeated several times over the next few years.

Tamper-proof head: There have been many proprietary fastener designs to provide a measure of protection against unauthorized removal. Street sign attachments and early consumer electronics products come to mind as having these fasteners installed. In this design, a cold-headed punch for each head size produced a head that was both oval and with tapered sides. Attempting to grip the head with locking plier results in the pliers slipping off easily. It was simple to cold head with the correct punches, which were closely controlled. It was a simple efficient solution to a common problem.

Fuel pump ball bearing seal: In the similar approach to assembly protection, a diesel engine fuel pump manufacturer set the fuel flow pump rate with a special button head socket screw. To prevent potential harmful tampering with this adjustment in an attempt to create more power from the diesel engine, this manufacturer drove a small ball bearing into the hex socket of each screw once installed and adjusted to the factory flow rate specification. It is a simple and effective solution.

Thread rolling machine die pressure adjusting screws: In a clever use of opposing screw thread forces, many flat die thread rolling machines use a set of four screws installed in the die block which holds off half of the thread rolling die set. While there are several variations of this design, the basic design has two screws installed next to each other and parallel. One of the screws will have a flat point, which presses against the die holder, moving it toward the blank being rolled, thereby filling out the threads more. The adjoining screw pulls the die block back away from the blank, reducing rolling pressure. One pair of screws adjusts the starting rolling pressure, while the other set of four adjusts the thread-rolling pressure at the end of the roll-threading pass. By working one screw against the other, both starting and finishing rolling pressure can be fine-tuned in surprisingly very fine increments and also locked into place during the vibration setup in the machine during thread-rolling operations.

Shear bolts: The fastener represents a clever solution to a commonly occurring problem. Acting as a mechanical circuit breaker, the shear bolt uses a reduced area section to failure in shear above a design load. The sheared bolt prevents a more catastrophic failure in a more valued part of the fastened assembly. The reduced area must be controlled dimensionally. It also is dependent on a specific range of metallurgical factors, which along with the shear design area determine shear load of any individual shear bolt under design operating conditions. As a teaching example, the shear bolts can be measured dimensionally and the diameter and width of the shear section dimensions plotted, say on an X Bar R diagram. They can then be analyzed statistically to determine shear dimension control and process variance. Samples of bolts can also be destructively tested with the test results studied for process control purposes.

Multiple start electrical connector screw and housing: Maybe it is because I served an apprenticeship as an electrician. An electrical connector with a set of three multiple start screws always impressed me by how quick the connector was to open up to make the wiring connections and then tighten up to put into service. These small-diameter screws have what I believe are three starts of a screw thread for forming into plastic. Spaced at 120° intervals around the connector assembly, each 360 rotation of the screw advances or removes the screw what appears to be a 6–8 mm. Neat design!

Disc brake caliper screw: Automobiles have always been large consumers of fasteners. The disc brake caliper screw I refer to here had a hex socket head drive, a section of machine screw thread from just below the under head fillet area to a long controlled diameter shear pin section. This long pin section aligned the two halves of the brake caliper body. It also provided support for the disc brake pads as they moved against and ways from the disc brake rotor in operations. Used in large quantities, the pin section was initially centerless ground to control the diameter. With continual process development, this sophisticated "special" screw became a production standard headed and threaded with minimal secondary operations. This application represents team work between the product designer, fastener manufacturers, and the assembly operations.

Sealing screw: This special screw has several features to commend it for our attention. To prevent moisture from entering an assembly, this pan head type of machine screw has an under-the-head area called a reentrant fillet. A fillet is a curved section which on a typical screw joins the head to the body of the screw. With a reentrant fillet, this curved section is completely in the underside of the head. They have been around for years and can be cold headed with a header die having a mating bump on the face of the die. They are not easy to run, and this bump feature on the header die face wears down quickly. I have heard the term "tool eater" referred to in these jobs. One of the things that makes the sealing screw technically interesting is that a polymer O-ring is placed in the under-head reentrant fillet. The O-ring dimensions are selected so that its cross-sectional area fits half in the fillet and half protruding below the face of the screw. The O-ring diameter is matched to the fillet diameter so that their circumferences are matched. When the screw is installed and tightened, the O-ring compresses and makes a water tight seal with the mating part surface. The second interesting point was the inspection gage used. A piece of clear industrial-grade polymer sheet was drilled and tapped. The sealing screws were required to thread into the clear plastic plate until seated with complete fill of the compressed O-ring with no unfilled areas visible or any excess O-ring material sheared off. It was a quick, efficient solution to a fastener application engineering problem.

Cold-headed socket wrench extension bars: Several areas have points of interest. The application involved the socket wrench extension bars found in most mechanics' tool sets. A potential customer called my employer's sales office. He requested a meeting to discuss cold heading some high-volume parts his company sold to a major hand tool supplier. I was asked to meet with this gentleman. I had been in this company's employment and had been working in the cold-headed fastener business not very longer than one year. The man brought a shoe box full of drop-forged extension bar blanks. He explained that with annual volumes of the hundreds of thousands of each sizes, drop forging was slow, costly, and inefficient. The hot-forged extension bar blanks required extensive metal finishing operations to remove the scale and flash inherent in a drop-forged part. He was motivated and convinced that these parts were good candidates for cold heading. There were three different square drive sizes and three different lengths for each drive size. The part had a head with a deep square socket, a shank of varying length, and a square tenon at the end opposite the head and square socket. The company was part of a large industrial conglomerate. The corporation had a cold-headed fastener division whose tooling engineers assured him that cold heading was technically not possible. It may have been this last statement that most focused my attention rather than the potential sales volume involved. I had been taught as a boy working with my family that there is no such thing as cannot. I did not have even a small amount of on-the-job cold-heading experience even though I had been working on machines and hot rods from my first second-hand bicycle through hot roads and Navy ships. I had also convinced my new employer to send me to a 3-day cold-heading technology workshop, where I gained a lot of cold-heading, tooling, and materials information. This was key to the eventual success. I did not know that it could not be done, and I had the benefit of some really good technical fundamentals of cold heading. I was new and had not acquired a lot of time-demanding projects, which would have motivated to take the safe route of saying "No". The next important point is that we, as a company, logged it in as a request for quote and I started to "do my homework." As mentioned under the cold-heading process, the three primary cold-heading operations are upsetting, forward extrusion, and backward extrusion. The extension bars would require all three. The head would need to be upset from the heading wire diameter, a square socket backward extruded into the head, and a square tenon to engage the various the various socket wrenches forward extruded. Fortunately, these volumes and forces can all be calculated to provide a good engineering estimate of feasibility. And I had been fortunate to learn the equations at the cold-heading workshop. The calculations showed that our existing headers were capable of heading these parts. We estimated the tooling, fixed, and variable costs as demonstrated in the engineering economics section. The company had a profit margin formula based on a "hurdle rate" formula, using the gross profit margin "hurdle rate" divided into total job cost. We prepared our quote and made our presentation to the customer. While our ability to attempt the job got their attention, our numbers did not. Not only did we need to provide a purchase price less than their existing in-house drop-forged cost, we were required to show a 30% savings to justify farming the work to an outside supplier. We worked at our costing and were able to offer that less finishing would be required on both the forward extruded square of our cold-headed parts and less over-preplate finishing of the entire part. Several rounds of quoting followed. The term "Using quotes as range finders" was expressed about our approach, and I think not unfairly. The potential cost savings was finally too great to not try and we were awarded a trial order. A learning point is that we were confident both in our engineering calculations and cost estimates, and in the potential revenue and profit of a successful completion of the trial order. Another is that almost everyone in the company "caught the fever" of this trial order. I have since coined the term for this phenomenon—"the Tom Sawyer effect." Everyone knew of this effort, most were intrigued by the technical aspects and their combined team energy and efforts carried the trial order over many tooling and header setup and tooling development obstacles. Those of us at the start of the project championed it company and supplier wide, but I'm convinced that the team support to make it a success is a factor in most new fastener and headed-part projects. This one worked, the customer was pleased, placing bigger and more frequent orders for increasing quantities of extension bars. Then their competitors wanted to have theirs cold headed, and we never looked back. An interesting side note is that the cold-heading machine manufacturer assured us at the start of the project that their machine could not cold head these extension bar shapes. Once the jobs were running on several of their cold headers in our shop, they requested that they be permitted to visit to observe their operation on the "can't be done" parts as they had several customers who wanted to buy the same machines. We took this as a compliment, and declined.

The "Tail Piece" and the "D"—shaped door striker: In the previous example, the extension bars were what we called headed parts, under the general product category specials, as opposed to standard catalog socket screws, which were graded A, B, and C depending on the frequency of their finished goods turnover. When the project started, the plant was backlogged with standards, with only "A" items being given manufacturing priority. Eighteen months later as the extension bar orders rapidly ramped up in order quantities, the economy entered a deep recession, and everyone was glad to have the work. I noticed an interesting behavior that I believe is common in fastener operations. At the start, most were cautious about the chances of success in heading the parts. Once it became an everyday operation, a mode of anything is possible on the headers and roll threaders took hold. We swung from cautious to overly confident, in hindsight. Fortunately, our corporation was in an acquisition mode and a small fastener shop on the opposite side of the country became available. The partners were growing the business and running against growth constraints in capital, among other things, and as with many founding owner operators, were ready to cash in on to some extent, on their hardwork. One of the industries they served was the residential door lock manufacturers. In addition to threaded fasteners, several complex steel parts were used. One was a rectangular shanked part called the tail piece. These owners figured out how to tool this part and more importantly, cold form it on their older headers. The innovative part was that to make it possible; they had the heading wire for this part drawn by their supplier in the shape of the rectangular shank less the clearance to be pushed into the rectangular header die by the first blow punch. The learning point is that they did not accept the limitation of trying to head round steel heading wire into a rectangular shape. They jump started the process by having the wire preshaped. A good application engineering mindset is to view the entire system, which in this case included the shape of the raw material. Following up on that clever piece of heading tool design and operation, each door has a striker bolt which is "D"-shaped and has a curved face. If you look at the side of any interior residential door you will see the end of this part. Using the saw-shaped wire feed into a much larger multiple die transfer cold header the corporation was able to provide funds to purchase, these clever operators leveraged the new capability and their earlier independent success to provide this much more complex, previous powered metal or cast part for a big savings to the customers and a nice increase in their sales and profits.

Repair screw: This one is simple, but also offers insights into good fastener application engineering. Many injection-molded plastic products use self-threading screw threaded into molded holes, often into bosses. These injection-molded holes typically have a tapered inside diameter resulting from the draft angle on the core pin(s) of the mold tooling. This core pin angle is necessary to allow the molded part to be ejected from the injection mold tooling at the completion of each mold cycle. It does decrease the holding strength of the metal screw with either a type "B" or perhaps a proprietary spaced thread form designed for mechanical fastening of plastic. In fact, this past week as I was writing this section I took apart a garden tool to repair an electrical connection that had come loose. I removed and reinstalled four of these screws in the repair. With each reuse, the holding strength is reduced as the hole site becomes increasingly "plowed up". I first saw the concept of repair screws in the consumer camera-manufacturing market. The small M1.4 to M2 diameter thread forming screws for plastics had a small run of repair screws made for just such reuse application. As an example, if the application was specifying M1.4 diameter thread forming screws for plastic, the repair screw for this application would have the same thread pitch but be M1.5 diameter. The screw with a 0.1-mm larger diameter can then tighten with an increased "bite" into the previously fastened hole. This provides the repaired assembly with increased joint integrity. The cost to produce a small lot of repair screws was minimal, requiring a slightly larger diameter roll thread diameter. Often worn heading dies from the high-volume M1.4 screw blank production run, which had developed a bell-mouthed condition from the hire cutoffs riding over the die bore entrance, could be lapped open to 0.1 mm larger, thus saving a worn-out die for productive use. The rolling dies were the same, set with additional clearance to roll up to full major diameter the slightly larger diameter repair screws. Some users even had the repair screws plated a different color to facilitate their ready identification. Although production screws were zinc and black chromate plated, the repair screws were zinc and clear. This same technique has been used with water-handling valves. The class 5 fit studs of high-usage pumps are sized to fit in the rebuilt vales, with the slightly larger-than-original valve internal threads being fitted with correspondingly larger diameter replacement screws and studs.

It is hoped that the above applications will provide ideas and inspiration for your challenging fastener and headed-part applications in the future.

Reference

1. Jim Speck, P.E., *Fastener Fundamentals Seminar, National Industrial Fastener Show and Conference*, Columbus, OH, May 1995.

4

Economic Factors in Fastener/ Assembly Decisions

Industrial commerce can be likened to a competitive professional sporting event. Like the competing teams, the industrial "players" are selected to be part of the "team." The team practices to strengthen its proficiency in the fundamentals of the sport. During the competition, the teams compete against each other for the goal. Only one usually wins the ultimate goal. After the contest, and a short time of celebration for the winner, the cycle repeats with team selection and practice for the next commencing season. During each new season, the competitive process usually results in new and improved skills and equipment. In fact, continuously improving skills and equipment are necessary just to stay at parity, let alone prevail.

Industrial products have their components designed and manufactured in a selection practice established to let them be successful in the competitive marketplace. They are selected for the assembly just as athletes are recruited for the team. Simulations, practices, and dry runs are performed to shake down the assembly process. In most industries, ranging from aircraft manufacture through consumer durables, electronics, and defense, there are key assembly skills and fundamentals that are specific to that industry, and the goal is nothing less than economic survival and prosperity. There is triumph for the successful—elimination from the market for the less so.

To succeed in the commercial arena, where competition can be fierce and unforgiving, your assembly must, at the least, meet the industrial parity of assembly economic efficiency, all other things being equal, if your product is to be successful. To prevail, and truly grow market share and achieve sales growth, your assembled product needs to be better assembled than your competitor's product. It must come together faster, with fewer pieces, and use less time, money, and energy in doing so. Seen in this way, the assembly and joining of a commercially successful product requires assembly knowledge and careful planning.

The assembly process can be thought of as a grid as shown in Figure 4.1.

As the assembly systems designer, you lay out the playing field when you plan and establish your assembly lines and processes. With these in place, products of your work flow through or across your assembly lines. Once the structure of the assembly and joining process is laid out and resources committed to it, with machinery and equipment in place, major changes in

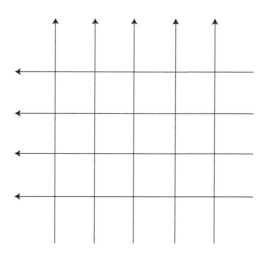

Components and fasteners

FIGURE 4.1
Assembly process model.

process require significant added costs, which can weaken your competitive standing in the marketplace.

Knowledge of your assembly goals, with a strong bias toward engineering and economic efficiency, is the foundation upon which to build your assembly and joining operations. Once in production, only fine tuning for continuous improvement will, hopefully, be all that is needed to keep your assemblies competitive and in demand.

On a technical level, the fundamental decisions—the how, what, where, and when of fastening and joining decisions—settle on how *strong*, with what reusability, and with what appearance. The earlier in the engineering design stage that these decisions are studied, the greater is the probability that their impact will be positive as a competitive factor. To understand the value of this early-on consideration of fastener economic factors, it will be useful for us to look at a hypothetical application where the opposite fastening design approach is taken. Suppose we have a new point-of-purchase vending machine coming on stream in our product line. Let us also suppose that we postpone any serious consideration of fastening and joining design, procedures and hardware selection, and purchasing sources until we have components arriving at the assembly site. Our assumption is that we will just buy loose hardware, namely screws, nuts, bolts, and washers, from whatever sources are readily available. As unrealistic as this may sound, it is in reality far more common in the real-world practice of industrial assembly than might be imagined.

Let us pick up our application in the assembly area. Purchasing in turn is pushing the local fastener distributors. We are calling for one after another

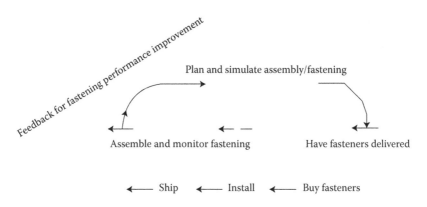

FIGURE 4.2
Two models for the assembly process.

bolt, nut, pin, screw, and washer type as we try to match what will fit our assembly with what is available. As individually shipped cartons arrive and are opened and inspected, our accounting department is subjected to a stream of invoices for fastener hardware, much of it at small lot and expedited delivery pricing.

In installing the fasteners, much time is consumed figuring out how to do the assembly work. Clearly, there has to be a better way to approach fastener fitting and fastening than with sometimes less than ideal hardware by trial and error. Since we do not have a clearly laid out plan along with diagnostic quality checks, our associates, employees, and customers cannot be assured that assembly forces are robust enough to return long-term service and durability. The most we can say, with relief, is that we got the job done. Definitely, there must be a better way to approach the fastening function.

Figure 4.2 shows two models for the above assembly process. The first model is pretty much how we described it, and again, happens all too often in industry. The second shows a more efficient approach, one which allows us to zero in on more efficient assembly and joining methods with fewer frazzled nerves in purchasing, engineering, and manufacturing.

4.1 Build Quantities

After our assemblies' target market is identified and the design objectives are well defined, it is strategically important that the build quantity be forecast. Realizing that build quantity forecasting, as with any futures-based science, involves a significant degree of uncertainty, it is still incumbent on all team members in an assembly project to have a clear understanding of how many assemblies may need to be built. A key factor in the fastening costs of an assembly are the number of assemblies that are planned for various units of time: per-hour, per-day, per-week, and per-year number of assemblies, and those over the planned production life. No doubt, projecting with accuracy the build rate and quantity presents a difficult challenge, especially for the assemblies of new product for which the demand is relatively unknown. Regardless, assembly team members, especially those with marketing insights and perspectives, need to research with precision and communicate how many and at what rate the assemblies will be put together.

There are a surprisingly large number of factors that are dependent on the build quantity planned and realized for most products' assemblies. An obvious one is the cost of the fasteners and joining components and compounds. Less obvious perhaps are the assembly tools, fixtures, and equipment that will be required, or the specific layout and arrangement of the assembly site. Another key factor is the structure and extent of assembly training that will need to be planned, documented, and carried out. And concurrently, the fastening materials scheduling, manufacturing, and distribution usually works best when established at the outset with informed involvement by everyone in the supply chain, from manufacturing to assembly, based on a foundation of reliable and up-to-date assembly build forecast information, with everyone working from the same page.

We can sum up the build quantity forecast factors as follows:

1. Fastener material cost
2. Tools, fixtures, and equipment
3. Assembly layout and arrangement
4. Assembly supplies distribution

Let us look at the components of these factors for ideas on how to plan and control them before we look at production.

Fastener material costs can be broken down into the cost of the raw materials used to make the fasteners. These could be steel, stainless steel, brass, aluminum, or a mixed range of these metals. They could also be plastic resins, usually of engineering grade. Or they could be chemical compounds. In anticipating the quantities of assemblies to be built, and the number of fasteners to be used in each assembly, the impact of the cost of the general fastener material on the cost of producing the assemblies can be figured into the cost of manufacturing those assemblies. In a more proactive way, by knowing the trends in fastener raw material prices, a company can anticipate the price change effects in raw material prices, which may affect assembly manufacturing costs.

An example of this would be the use of 300 series, 18-8 stainless-steel fasteners in a proposed new assembly. We can increase the corrosion resistance of our fastenings and eliminate heat treating and plating costs. A prime cost driver of 18-8 stainless-steel fasteners is the nickel used in its composition. For example, UNS 30430, commonly known in North America as type 302 stainless, contains nominally 8% nickel. By recognizing the influence that nickel cost has on the cost of stainless steel cold heading wire, we can project the impact that increases, or to a lesser extent decreases, of nickel ore have on the purchasing costs of our stainless steel fasteners. By multiplying this out through the forecast assembly build quantity, we will know how many dollars of assembly cost are dependent on, and leveraged by, stainless raw material and their fundamental nickel prices, domestically and internationally. Taking a proactive stance, a company contemplating a fastening system requiring a degree of corrosion protection or resistance could compare stainless fastener raw materials with other fastener raw materials for short and long-term cost variation factors as they impact on the forecast product's cost to assemble.

Similarly, petrochemical raw material-based fasteners such as those made from engineering-grade plastics can be economically impacted by large fluctuations in oil and, thereby, resin prices.

Build quantity also greatly influences the manufacturing decision regarding the fasteners that are used and the manner and timing with which they are obtained. If quantities are relatively small, compared to the production method that is standard for those types of fasteners or fastening systems, then purchasing quantities will be obtained from portions taken from larger fastener production runs. Suppliers' early involvement can highlight the particular supply history for candidate fasteners and make recommendations on ordering quantities and frequencies, which can ease procurement problems in the future and also help obtain the most economical cost for the fasteners.

If the assembly build-rate forecast results in production quantities that are larger, to an amount approaching the normal production runs for the fasteners, close communication between everyone along the supply channel, from the fastener application site back to the fastener production source, should proceed as openly as possible so that fastener production efficiency will mesh with assembly fastener consumption rates.

Continuing on with the influence of assembly build quantities and fastener economics, a review of the specified fasteners' manufacturing processes can often show at least several steps that can be tuned to the assembly build quantities. Machine setup and production rates have amortization factors. Batch processes such as heat treatment and plating and finishing often have minimum lot limits. It is economically beneficial to match the required fastener consumption to fastener production methods. This can offer an especially significant economic advantage to the proposed assembly if candidate fastener decisions can consider each choice at production rates in comparison to assembly build rates prior to laying out the final assembly drawings and bills of materials.

Consider two designs for a special retaining pin designed for manufacture from low-carbon steel. The pin can be either screw machined or cold headed.

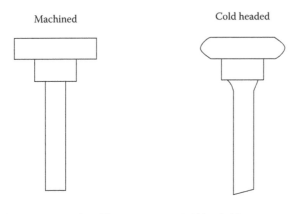

Machined   Cold headed

Machined from round steel bars   Cold headed from wire

Well-defined corners and close   High-speed production, stronger grain flow
tolerance control but low material   but corners radiused and less tolerance
yield and slow production speed   control

FIGURE 4.3
Retaining pin process comparision.

The screw machine part has a lower setup cost for getting us into production. Due to its processing rate and material consumption, it also has a higher per-unit production cost. The cold-headed version has higher start-up costs for tooling and development, but once over a critical volume offers lower unit costs. Figure 4.3 shows both versions of this fastener.

4.1.1 Tools, Fixtures, and Equipment

The assembly build rate also has a key effect on the type and extent of the assembly fixtures, fastener installation tools, and the joining equipment, which will produce the most efficient assembly procedures. The initial decisions on tools, fixtures, and equipment can also have long-term effects of both an economic and product integrity nature. Well-planned installation support equipment that is matched well with the demands placed on it by the assembly build quantity requirements will perform efficiently in installing the chosen fasteners and making the assembled components secure. If well matched to an accurately forecast build rate, it will neither be underutilized and be an underperforming asset nor will it be one pushed beyond its capacity by a larger than planned demand, resulting in fastening equipment breakage, downtime, and, perhaps, underfastened joints, which may fail when used by the assemblies' customers.

Tools and equipment can include the following:

1. Wrenches
2. Electric, pneumatic, and hydraulic drivers

3. Jigs, fixtures, and clamps for positioning components and easing assembly ergonomics

4. Utility supply lines

5. Part feeders, vibratory hopper bowls and tracks, inspection gages, and machinery for delivering components and fasteners to the assembly site

6. Automatic installation equipment and robots

7. Numeric and computer controllers for programming fastening installation procedures

By forecasting the number and rate at which assemblies are going to be fastened and joined, equipment choices can be made to do the assembly work. In this way, economic efficiency is achieved for the assembly manufacturer while delivering a robust and durable product for the consumer.

4.1.2 Assembly Site

It should be intuitively obvious that adequate planning for the volume of assemblies to be built is a critical factor both to the seamless flow of components and fastening products through the assembly process and to the ergonomic safety, health, and comfort of the people operating the assembly processes and equipment. That this is not always the case should come as no surprise to anyone with even a passing knowledge of assembly sites and conditions. To those interested in assembly advancements, it should also be reassuring that the continuous improvements being made in fastening work have resulted in assembly layout, which make fastening work safer and more user-friendly for every person associated with its operation and maintenance.

First and foremost, an adequately planned assembly site will have sufficient room for the flow of materials and the safe movement of people around the assembly operations. If the assembly processes are configured as a line, they can consume a considerable amount of space. Good practice can include nonlinear but still continuous, free-flowing people, materials, and assemblies. Moreover, well-planned assembly sites do not necessarily require new construction. It does require an accurate forecast of the physical size of the materials, their rate of processing, the number of operations along with the nature of their assembly equipment, and the size of the operating crew and their range of motions.

4.1.3 Assembly Documentation and Training

If build rates are modest, and to be conducted for a shortened time period, simple verbal instructions may be sufficient. However, in many applications, the fasteners, along with the fastening tools, fixtures, and equipment, should be operated in a planned, timed sequence. This fastening sequence should

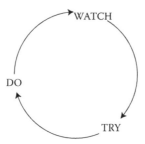

FIGURE 4.4
Training model.

be designed and proved for the forecast build rate prior to full production operation by manufacturing.

The key to this phase of assembly operation is fastening documentation and training. Depending on the nature of the assembly operation, clear, concise graphics and text should demonstrate the designed assembly procedures. All authorized assembly personnel should have access to and be familiar with the assembly documentation.

For training operators in performing the assembly operations, a useful training model is WATCH–TRY–DO. Figure 4.4 shows a model of this training technique.

4.1.4 Assembly Fastening Material Supply Channels

As a final note, a valuable step in good closure of the build-rate forecast and subsequent assembly planning is the cultivation and development of industrial fastening material distributors and manufacturers. They can bring to the planned assembly operations knowledge and experience that can compress the learning time required to ramp up assembly rates and offer solutions to build-rate problems as they occur. Their depth of industry contacts can also prove useful in tapping into additional resources as they are required, which can cut through red tape and speed up needed information and materials when critical situations arise. And the most effective time to start establishing this network is well before it is needed.

4.2 Assembly Robustness

At one time, the term "planned obsolescence" was used, with a somewhat negative connotation, to describe products that had a limited life. Often, they would be replaced at about the time they had reached an economic payback point. A classic example is an automobile, with an owner needing to replace it with a new one right about the time the last payment is made. In more recent times, many products are replaced for reasons of convenience and style rather than product failure. To the extent that planned obsolescence was a product strategy, competition and market forces limited its effectiveness.

Any assembly you design, manufacture, and offer in the marketplace for your industry will enter into competition for the buyer's funds. The durability of your assembly and the robustness of its operation cannot be surpassed by a significant margin without your suffering from a loss of market share. Returning to our automotive example, during a period when North American automobile manufacturers were perceived as having less durable products than other countries' products, their share of market and business fortunes declined. Moreover, it is important to note that the perception of reduced durability precipitated the loss. Buyers made value-weighted assessment of the market's offerings and purchased accordingly. It is important to consider an assembly's robustness and its perception. The factors which influence how robust an assembly and how durable its operation will be will both prove and be perceived as functions of the number of joint interfaces and the quality of their joining. An anvil is a very robust product in that it has no joint interfaces. In most of our products, the number of joint faces will be greater than this. As a designer, the goal should be to accomplish the desired assembly functions with the economic minimum number of parts. A good way to think of assembly reliability is as an inverse function of the number of parts. This can be drawn up as a graph showing the relationship between the probability of assembly reliability and a given assembly's parts count.

Pioneering work in the field of assembly design for manufacturing has been done by research, which emphasizes designing for assembly and has been made available as computer software programs and assembly models. By using design for assembly principles, an assembly designer can take a critical look at competing options for manufacturing an assembly and specify the design and parts that maximize robustness, durability, and value for the market competitiveness of the product, its ultimate consumers, and its manufacturer.

A key tenet of the less-is-more approach is to spreadsheet the entire bill of materials for an assembly. By laying into the spread sheet all of the times and costs associated with each individual part, its total investment by the assembly manufacturer can be compared with its function. In this way, many parts can be combined to reduce the total number of parts, reduce the assembly steps, and increase reliability while reducing assembly time and cost.

A caveat to using this valuable approach is that it is not a panacea. Several factors need to be considered by the assembly designer if the assembled components are to be successful. A key factor is the resulting complexity of the remaining fewer components. Since they may be performing more functions in the assembly, their design, sourcing and manufacture require early involvement by the candidate suppliers to better plan for feasibility. A good example would be self-clinching fasteners in sheet metal stampings. By using a self-clinch stud rather than a screw, washer, and nut, a design could reduce its parts count and assembly time considerably. It is incumbent on the designer and assembly manufacturer to ascertain that the self-clinching stud is readily available, what changes in the stamping (such as hole size and edge condition) need to be made, and to obtain by testing and/or research that the assembly provides the required assembly function and is robust.

And finally, it is important to consider the economics of the market for the intended assembly and the extent of its manufacture. It serves little purpose to be designing and manufacturing anvils if the market is moving toward adhesives. By knowing the goal of the assembled product, operating to deliver "head of the class" robustness and reliability, maintaining an early and continuous team approach to assembly design and operation, and quantifying each component's contribution and cost to the assembly's reliability, assembly value can be achieved. And with a diligent effort, it can be continuously improved.

4.3 Legal Considerations and Constraints

It is a cliché to observe that we operate in a litigious environment. It also serves little purpose to complain about what should be obvious to every assembly designer and manufacturer about performing their tasks with diligence with regard to the legal exposure their work can entail.

The Latin phrase caveat emptor, "let the buyer beware," has evolved in the U.S. product liability environment to "let the manufacturer exercise care." It is important that both new assembly designs and those that are in production be reviewed periodically to determine any potential product liability exposure. In this way, steps can be taken, if not to eliminate product liability exposure to at least minimize its effects.

While it is beyond the scope of this book to provide an in-depth grounding in assembly product liability, several key concepts can be laid out for the reader. An important step is to understand, to the fullest extent possible, the nature of the intended market. The assembly designer and manufacturer should thoroughly research the use conventions and the product applications of the universe of their customers and prospects. A classic case of an irrigation pump design in the United States for agricultural use comes to mind. When it was sold for export use, it caused numerous field breakdowns and complaints. Only after much frustration for both buyers and sellers of these pumps was it realized that the new customers had little experience with powered equipment and were using the same relatively crude assembly and operating methods they had previously used with hand-operated pumps. Two-way communication can eliminate many fastener and assembly application problems.

Another important step is to anticipate, as best as possible, the manner in which a product or assembly can be operated incorrectly. If this seems far-fetched, consider the real-life case of two men trimming a hedge by holding a running gasoline powered lawn mower, blades spinning, between them with their hands on the four wheels as they walked with the hedge between them and the running mover cutting the top! No doubt today's dead man throttles and brakes are the results of accidents that arose from some of these product abuses. And again, it serves no purpose to rail against any injustice in cases of obvious self-negligence. A prudent manufacturer will anticipate, to the best of their ability, unsafe product use and design with this in mind. As a follow-up on product and assembly misapplication, a useful technique is something called failure modes and effects analysis, or FEMA, also known as F.E.M.A. By considering individual parts of any assembly and both the individual and any cumulative, domino-effect failures, an assembly can be made safer for consumer use by both strengthening individual parts and components and designing in redundancies for additional safety.

Also important from a diligence perspective is knowledge of the state of the assembly practice and fastening advances in your specific industry and product segment. Usually, no design is static. Most designs usually improve both from advances in the state of assembly practice of their competitors and from the judicial and legislative events in their industry. It is prudent for every designer and manufacturer to keep current on the legal proceedings of others in their markets and companies assembling products. A convenient and efficient method for doing this is to talk with your competitor's customers. Find out what represents best-in-class assembly performance. To the extent possible, understand their approach to assembly and compare your assembly performance to your best competitor's to benchmark the industry standards for fastening safety and excellence.

Perhaps, most importantly, you need to be as completely familiar as possible with your industries federal, state, and local laws concerning product safety and performance. The fastener industry uses several consensus standards organizations in the United States. Among them are the American National Standards Institute (commonly known as ANSI), the American Society for Testing and Materials (ASTM), the Society of Automotive Engineers (SAE), and the Industrial Fastener Institute (IFI). All of these organizations offer fastener standards, which can be useful in assuring the production of assembly work is performed with products and procedures that meet the standards of these organizations.

Addresses for these companies are as follows:

ANSI	SAE
345 East 47th Street	400 Commonwealth
New York, NY 10017	Warrendale, PA 15096
ASTM	**IFI**
100 Barr Harbor Drive	1717 East street
West Conshohocken, PA 19428	Cleveland, OH 44114

4.3.1 Consensus Standards

It is important to observe here that all of the above standard organizations produce technical standards that are of a consensus nature in the United States. Again, it is outside the scope of this book and author to offer legal advice. It is critical to understand that while in the United States, technical industrial standards, including those for fasteners, are developed by interested users, manufacturers, and technical people and are published for the benefit of all interested parties, it is the purchase order for fasteners that constitutes a legal contract between the seller and buyer of fasteners. For that reason, the language and specifications which are to be met should be clearly identified in the fastener purchase order. Many times, disputes over the performance of delivered fasteners can be traced back to haste and a lack of understanding of requirements and applicable technical standards on the part of both parties.

For this reason, a clear statement of the requirements of the buyer and installer of the fasteners, fastener tools, and equipment in the purchase order is a good step in establishing the environment for a satisfactory meeting of requirements. And using consensus standards as a purchase order callout can simplify the language needed to communicate the requirements. By using a known standard, both buyer and seller can be "working from the same page."

As a follow-up to the use of established fastener standards, a useful process to employ to assure industry standard performance of purchased fasteners is the use of accredited fastener laboratories to carry out the tests. There is, as of this writing, one accreditation body, A2LA, which has audited and accredited a wide range of fastener and metal laboratories throughout North America. The National Institute of Standards and Technology (NIST) also has prepared a fastener laboratory program. The elements which constitute the accreditation of a fastener laboratory are a documented system of procedures for each type of test performed. Also, proficiency testing is performed to known standard or certified reference materials. Adequate training of laboratory personnel is carried out and documented. And finally, periodic auditing to International Standards Organization (ISO)-based auditing procedures is performed to reaffirm the laboratories' accreditation.

By using an accredited fastener laboratory to periodically test or audit fastener performance on application-important fastener properties, assembly confidence can be increased.

4.3.2 Some Fastener Cases

4.3.2.1 Jet Engine Mounting Bolt

One fastener application that was widely reported involved the engine pylon bolts of a wide-body airliner. As this large jet aircraft, filled with passengers, took off from Chicago's O'Hare airport, an engine literally fell off the plane. The jet crashed, with a tragic loss of life and damage.

During the subsequent investigation of the crash site, an aerospace quality, 12-point drive bolt, complete with manufacturer's identifying head marking was discovered by the search team. It had not fractured and speculation was reported concerning the possible loosening of this bolt as the cause of the structural engine mounting falling off.

The fastener manufacturer's president was quick to acknowledge that they were indeed the bolt's supplier. By thorough investigation, it was ultimately established that the bolt had neither suffered a failure in its integrity nor loosened and vibrated out on its own. It was discovered that the plane's engines were serviced incorrectly by using a fork truck to lift up the engines while working on them. This strained the triangular-shaped pylons in ways they were not designed for, thereby precipitating micro-sized cracks in the pylons. These cracks then vibrated and grew during takeoffs and landings until failure occurred.

It was due in part to the bolt manufacturer's rigorous production testing and documentation that gave the investigation sufficient data to eliminate the bolt as the failure cause and investigate further. By doing so, they were able to discover the incorrect maintenance procedure and stop it to avoid further tragic occurrences and also lead to a redesign and strengthening of the pylon.

4.3.2.2 Propeller Screw

Another aviation application involved the fracture of some special socket head cap screws used in the propeller hubs of variable pitch propellers on commuter aircraft. The screws and application had been in production for a great number of years with no experience of failure. The propeller manufacturer and fastener producer were both from North America. The failure occurred in the mountains of South America. Numerous fasteners from existing and past lots were mechanically tested in the fastener producer's laboratory by their metallurgist. Also, records from previous years' production were examined. One of the tests conducted was a fatigue test run over a wide range of loading spectrums. By designing the fatigue tests to determine which load levels and frequencies could cause failures in the number of cycles indicated by the plane propeller's age, it was determined that it would be loading greater than originally designed for with the screw in question. When this information was transmitted to the aircraft's operator, it was reported back that indeed, the aircraft's engine had been uprated with a significant increase in power to enable it to fly in a high mountain environment. Subsequently, this new loading data was used to engineer a redesign of the screw to handle the increased loading with an appropriate design factor.

4.3.2.3 Motorcycle Disk Brake Rotor Screws

While in the previous two cases, failures occurred which were not attributable to the fasteners, a motorcycle manufacturer made a change to a front disc brake design on one of its most popular models and experienced fracturing of the fasteners right in the plant as the wheels and newly installed disc brake rotor assemblies traveled through the plant to be assembled into completed motorcycles.

An investigation by the author led to the fact that failure of the disc brake rotor screws was being caused by hydrogen embrittlement of the screws during finishing, with the probability being the electroplating process. At the time the author's company was not the fastener's supplier, so detailed investigation of production processes was not an option.

The embrittled lots of screws were rejected from production use. New fasteners were produced. As an incoming process step, it was decided with the customer that a statistically significant sample of each incoming lot of disc brake rotor bolts would be placed in test fixtures and given a Mil Std 1312 stress durability test for 200 h. Only when the lot samples passed the test successfully would the lots be accepted for production use. The results were that the in-plant failures stopped. Also, testing revealed which lots failed prior to their being accepted and the fastener manufacturer then had data from which to track down and eliminate the variable(s) causing the embrittlement.

4.3.2.4 Fractured Valve Screws

A gas mixture valve failed in service and resulted in a lawsuit. During the litigation, the valve manufacturer sued the fastener distributor. A broken screw had been found in the failed valve and a claim was made that this screw was defective and had led to the valve failure. The author was brought in as an expert to help determine the cause.

A new valve was obtained as well as typical examples from the distributor's inventory of the screw. It was a fairly common type "F" self-tapping screw, which was used to self-tap into the valve's die cast housing and secure a smaller control valve inside the main gas valve's body.

Some examinations were conducted to determine the possible causes of the broken screw found at the failure site. Tests were conducted to determine the torque and resulting friction developed during installation. Engineering analysis was also performed to determine the possible service loading.

In the end, several factors were identified which could have caused the failure. Some were related to the installation of the screw, whose placement required some unergonomic bending of the hand and screwdriver to access the screw's cross recess drive. No final determination could be made as to the cause of failure, but factors were identified which could hopefully prevent future occurrences.

4.3.3 Bogus Bolting

A fastening situation that received much national attention, and whose consequences have not to date been completely resolved, involved what has been called the "bogus bolt" controversy. At the root of the controversy was a distributor who decided to investigate how alloy steel grade 8 hex head bolts could be sold so cheaply in his marketing area. He suspected that since the prices were much lower than could reasonably be expected, given their material cost, content, and processing, something did not seem right.

A grade 8 bolt, by consensus standard and convention, has 6 radial marks on the face of its head. These indicate a grade 8 tensile strength and corresponding approximate Rockwell hardness. In checking the hardness of these *very* inexpensive grade 8 bolts, their hardness tested within the specified hardness limits. It was only after a detailed spectrograph analysis of these bolts' metallurgy that it was found that they were not manufactured from alloy steel at all. A much less expensive low-carbon steel, with a small percentage of the element boron, was used to manufacture some imported lots of grade 8 bolts. It has been known by metallurgists since the 1940s that trace elements of boron, with as small a material composition percentage as 0.05%, will produce a hardness of the same level as a midcarbon alloy steel. Using a steel such as 10B30 would result in the same HRc 36 hardness reading when the bolts were hardness tested by an inspector. And of course, they were much less expensive to cold head due to the much lower raw material cost.

They are much lower in important mechanical properties such as impact strength, ductility, and fatigue strength. They were, and are, much less able to maintain strength at the slightly elevated or reduced temperatures which grade 8 bolts are by design capable of withstanding.

The fact that they were marked and sold as grade 8 was pure fraud. It is unclear how widespread this practice was or its extent. It is noteworthy that when the Federal Government started to investigate, several agencies, including the FBI, did find and prosecute and in some cases found outright fraud, as in the case of the grade 8 bolts mentioned previously as well as military specification (MS) fasteners whose procurement specifications require specific testing and documentation. Also found and generally improved was some carelessness in the procurement, inspection, and documentation of fasteners for government use. On November 16, 1990, Public Law 101-592 covering ¼ in. or M5 or larger diameter fasteners which were through hardened was signed into law. Its implementation has been withheld pending amendment to make some sections of the law less onerous on honest manufacturers and suppliers of fasteners. It is worth listing, for reference, the legal considerations of fasteners and some fastener events having legal ramifications, courtesy of Fastener Technology International.

September 9, 1979—Correspondence from the Bureau of Consumer Protection to the Federal Trade Commission stating: "The tragic loss of 274 lives in recent DC-10 air crash prompted this office to investigate allegations that counterfeit or otherwise materially altered and unsafe air craft fasteners are being sold and used by commercial airlines in the United States."

May 9, 1988—Reports that problem fasteners were found in large numbers in the vehicles of the Seventh Infantry Division at Fort Ord, CA, and Ninth Regiment at Fort Carson, CO.

May 10, 1988—U.S. Army told Congress it scrapped more than 30 million bad bolts over an 8-month period and that an unknown number of these bad bolts still remained in its weapons where they can work loose and cripple weapons and soldiers. It has also stated that tests conducted on the previous year's inventory revealed 30% of the common bolt inventory fell short of requirements.

June of 1988—The *Commercial Carrier Journal* published a 10-page article relative to the discovery of counterfeit bolts in truck 5th wheel installations and other critical truck and bus components.

June 9, 1988—A Nuclear Regulatory Commission official told the U.S. House Subcommittee that more than half of the nation's 109 nuclear reactors had substandard bolts in safety-related locations.

June 10, 1988—U.S. District Court of California issued a search warrant to fastener supplier based on falsified test results on bolts used on the Trident II Missile.

July 31, 1998—Substandard or counterfeit metal fasteners were linked to the death of an ironworker from Tennessee working on a U.S. highway bridge in Louisiana.

August 26, 1988—A government aircraft alert issued on numerous fasteners purchased from fastener supplier by a major aircraft builder. The alert is based on alterations and mismarking of fasteners.

September 26, 1988—Fastener supplier charged with 26 counts of false statements and 17 counts of mail fraud. On 11/30/88 subject pleaded guilty to 43 counts of fraud and false statements. On 12/12/88 subject was put out of business and fined $62,150 plus $34,500 reimbursement fee.

November 1988—During erection of a 230 kV lattice tower, eight 3/4" diameter bolts broke. Subsequent test determined 20% of the lot failed due to lack of stress relief, shear bands, and zinc migration into bolt surface.

November 5, 1988—Inspector General's review of 685,000 parts in Georgia firm reported that 90% of the parts were substandard, failing to meet specifications, or useless. Among the faulty parts were bolts used for the tail drive section of H-3 helicopters.

January 27, 1989—West Coast newspaper reported several people lost their lives in crashes involving private planes that officials determined were caused by defective fasteners.

February 18, 1989—NASA impounded thousands of bolts and examined every fastener on the space shuttle after inspectors discovered that manufacturers were faking certifications.

February 20, 1989—Twenty federal agents seized 52 crates of documents, test equipment, and fasteners in raid on firm in which fictitious inspector ploy was uncovered. Bolts were to be used for the B-2 Project.

May 13, 1989—Canadian defect investigation on death of tractor/trailer driver pointed to pinch bolts, failure causing detachment and death. The bolt hardness was out of specification and in a brittle condition.

June 27, 1989—U.S. District Court-Northern District of Texas filed charges against 12 companies and individuals over a "scheme and artifice to defraud." Eighty-seven tons of suspected goods were seized.

July 22, 1989—Federal investigators were studying the possibility that a dislodged nut may have been sucked into the rear engine of a commercial plane causing the engine to fail. The aircraft lost control and fell in a fiery crash.

August 10, 1989—Jet engine builder offered $279,000 in rewards to Iowa farmers who may have found missing aircraft parts from a DC-10. The tail engine blew apart and 111 lost their lives.

August 13, 1989—Major U.S. retailer recalled bolts from faulty swing sets that could toss children to the ground.

September 27, 1989—Release of Defense Criminal Investigation Service Report (dated July 11, 1989) that the Pentagon was auctioning off scrap bolts sent to junk dealers who resold them for use in commercial aircraft and military systems.

October 2, 1989—Chicago newscaster reported infiltration of bogus fasteners into U.S. military was widespread and epidemic.

November 1, 1989—Letter sent from Congressman to Subcommittee on Readiness that 750,000 fasteners in electrical switching boxes that connected 1100 MX and Minuteman ICBM launching mechanisms did not meet specs.

December 18, 1989—DOD Inspector General investigation precipitated $2.8 million penalty on firm for false marking and invoicing boron steel fasteners.

December 18, 1989—National Highway Traffic Safety Administration (NHTSA) notified its regional associates that certain suppliers of Grade 5 and Grade 8 bolt head markings did not meet SAE J-429 and ASTM A325 standards for such bolts. Associates strongly urged to demand certification reports and perform periodic inventory audits of fastener stock.

December 22, 1989—Company pleaded guilty to five felony charges out of office of U.S. Attorney—Northern District of CA, for falsely certifying aerospace fasteners that were sold to the U.S. military aerospace programs.

February 21, 1990—Sheared bolts on cantilever-type road sign blamed for death of 41-year-old woman in Michigan.

March 22, 1990—Company cited for substituting imported nuts and bolts for U.S.-made products in highway guard rail application.

May 7, 1990—Fleet of CH-47D Chinook helicopters grounded after cracks were discovered in lot of barrel nuts used on helicopters.

June 13, 1990—British Airways pilot was sucked out of his cockpit when a windshield blew out due to 84 of the 90 bolts holding it in place being undersized.

July 30, 1990—U.S. Customs Commissioner testifying before Subcommittee on Oversight and Investigations Committee on Energy and Commerce stated: "Billions of substandard, mismarked, and/or counterfeit fasteners threaten the reliability of industrial and consumer products and our national security." Defective fasteners are not only a waste of money, but may in some case contribute to personal injury or death. The infiltration of substandard fasteners is due mainly to the profit incentive and deliberate evasion of standards upon which manufacturing procedures and product quality assurances are based.

November 26, 1990—Four bolts failed in a pump engine application causing a fire that burned 800,000 gallons of jet fuel at an international airport. Six hundred fire fighters expended 55 h to extinguish the fire.

March 9, 1991—Failed propeller pin caused crash in Key West killing three men.

September 3, 1991—West Coast newspaper reported that the Stealth Bomber program was beset by production problems including assemblies using wrong bolts and threaded fasteners.

November 15, 1991—Cargo plane manufacturer executives accused of approving the installation of substandard rivets on the wings of the planes.

June 1, 1992—U.S. Nuclear Regulatory Commission Notice 92-42 stated: "Fraudulent bolts in seismically designed walls [revealed] that heads cut from bolts were attached to the angle iron to make it appear there were bolts supporting the walls."

June 27, 1992—NHTSA received letter from a manufacturer of commercial trucks that wheel mounting studs on certain iron front and rear hubs may break and cause the tire and wheel assembly to separate from the vehicle.

July 7, 1992—Transportation Inspector reported "among 220 cases under investigation nationwide, agents have found counterfeit engine components, brake pads, thousands of low-quality bolts, and even junked parts that were welded and painted to look like new."

July 16, 1992—Water reclamation facility broke down due to shearing of anchor bolt clamps. Pipes were found strewn about in four of the tank's chambers.

September 5, 1992—Bolt jammed tether reel of satellite system in astronaut deployment exercise.

November 10, 1992—Officers of bolt supplying company plead guilty to selling Japanese-made nuts and bolts to federal contractor.

October 5, 1993—East Coast company subject of Civil Forfeiture Action seeking $2.2 million, luxury automobiles, for supplying substandard fasteners used in aircraft carrier, Titan missiles, and ground support systems.

February 11, 1994—Eight firms charged by government officials for pawning off low-grade items on the military.

September 14, 1994—Automotive company recalled 220,000 vehicles for fastener problems in brake assembly.

October of 1994—1994 utility trucks recalled for trailer hitch bolt problems.

January 6, 1995—Major defense supplier pleaded guilty to false testing charges and agreed to pay $18.5 million fine for selling potentially hazardous parts to the Pentagon. Substandard parts used on F/A carrier-based jets. About 1600 planes were involved.

As can be seen by the preceding listings, understanding of applicable standards by all parties involved their clear communication in purchasing transactions and compliance with the standards can be critically important from both a safety and a legal viewpoint.

As a follow-up observation on fastener issues in the public arena, while writing this edition, several applications have received attention due to fastener performance. A tunnel on the east coast had ceiling panel loosening and ceiling material striking the automobile of a passing motorist. On the west coast, alloy steel fasteners were found to have developed hydrogen embrittlement, causing bridge delays and costs for remediation.

4.4 Economic Factors in Fastener/Assembly Decisions

The economic benefits and costs associated with assembling a product are as important as the engineering mechanics of the fastener's performance. All other things being equal, an assembly which costs less to produce will outperform an equal but more expensive assembly. In analyzing fastener and assembly economic factors, we can view the fastening from a manufacturer, distributor, and consumer vantage points. Having a set of engineering economic tools will enable us to better perform this economic analysis.

A good first economic tool is the concept of fixed and variable costs. Fixed costs are the infrastructure that makes a manufacturing or assembly process possible. Variable costs are the costs that increase and decrease with the rise and fall with the number of assemblies being fastened. Both the fastening manufacturer and the fastening consumer's costs rise and fall with this change in fastening quantity. Figure 4.5 shows this graphically.

To help the reader gain a clearer understanding of the costs associated with fastener manufacturing, we will start with the manufacturing side of the process. For our example, let us select an M12 × 1 × 125-mm long hex bolt with flange-type hex nut. The bolt and nut will be manufactured in a cold-heading plant with both screw- and nut-forming and threading equipment, as well as equipment to heat treat and plate the screws and nuts prior to shipment. What does the plant general manager, sales department, and accounting department personnel see when they review the costs to supply the customer on the purchase order for these fasteners? First are their fixed costs. The real estate on which the fastener manufacturing plant sits has an annual cost. This could be the property taxes, amortized purchase costs, grounds maintenance, and insurance. Similarly, the plant building(s) have rent or mortgage costs, or amortizations. Property, fire, and casualty insurance costs are involved as are building maintenance costs. Utility costs such as heat, cooling, light, and power are factors. Compressed air for plant operations are partially fixed costs and partially variable costs. Anyone who

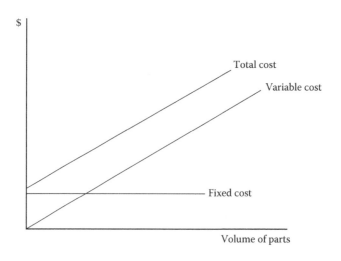

FIGURE 4.5
Fastener cost plot.

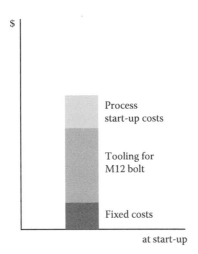

Process
start-up costs

Tooling for
M12 bolt

Fixed costs

at start-up

FIGURE 4.6
Fastener manufacturing cost breakdown.

believes air is free has never reviewed the costs of a fastener factory's air compressor systems. Furniture, fixtures, communication equipment, computers, and servers are part of the fastener factory's fixed costs. So, before we make fastener 1 for our M12 bolt and nut purchase order, we will have costs that must be paid each month. A graphic of this would be as illustrated in Figure 4.6.

Turning our attention to the variable costs, we can break them down into raw material, labor, and tooling. Raw material is the cost of material used to manufacture our order. In actual practice, we may be manufacturing these components with raw material from a much larger raw material purchase order. If the raw material required is relatively standard, a more economical cost might be obtained by purchasing it in larger quantities. Let us state that we are using 4037 alloy steel for both the nuts and bolts. Further, we will make our order quantity 10K parts. How much raw material is needed? First we need to know the volume, in cubic inches, for each part. Then we will multiply by a quantity factor and also factor in some overage, or scrap factor. To calculate the volume of the nut and bolt, we will calculate them as shown in Figure 4.7.

As can be seen, we use a 4037 alloy steel wire diameter at a diameter of about 0.001 inches smaller than the roll thread diameter to enable it to be loaded at production speed into the cold heading die. This is cut-off from a coil of several hundred, if not thousand, pounds. The length of wire is of sufficient volume to form the hex head when cold headed. For this part, let us state that it is 7.400 in. Our volume for the bolt is then this length times the area of 0.738 in.^{2}, which is 3.17 in.^{3}. Per thousand bolt blanks, this is 3165.5 in.^{3}. The weight of our wire is the density of 4037

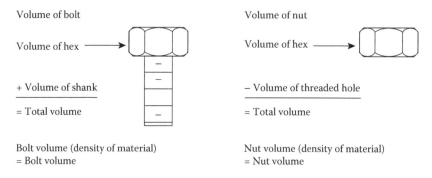

Volume of bolt		Volume of nut
Volume of hex ⟶		Volume of hex ⟶
+ Volume of shank		− Volume of threaded hole
= Total volume		= Total volume

Bolt volume (density of material)   Nut volume (density of material)
= Bolt volume   = Nut volume

To both, add a factor for scrap
2%–5% would be a common range

FIGURE 4.7
Fastner volumes with scrap factor.

alloy steel, or 0.29 pounds per cubic inches. Each bolt blank weighs 0.29 pounds per cubic inches times 3.17 in.³, or 0.9 pounds each. There will be approximately 90 pounds per hundred bolts and the entire order of 10K will contain 900 pounds. During manufacturing, we will lose some, hopefully small, percentage of wire as scrap. Examples are the short coil ends that cannot be cold headed into fastener blanks, blanks not completely up to dimensions during tooling setup and misformed from tooling misfires, perhaps from kinks in the wire causing "short feeds." These would be normal process variances of "scrap". The resulting scrap metal can be sold to a metals dealer, but usually only for cents on the dollar, depending on the economy and worldwide supply and demand for the metal. Special causes of scrap would be if the process drifts without detection by the operator and dimensionally non-conforming blanks are headed. Clearly we do not want the customer to have these. In avoiding this risk to the consumer, we do increase the risk to ourselves, the producer. This balance of consumer and producer risk is a fundamental of the process. If sort or rework cannot make them conform to specification, these then are also scrap. All scrap is inefficient and a waste. It is nonprofitable use of raw material. Clearly, zero scrap is the goal of manufacturing. Costing at a 5% scrap rate was built into the manufacturing cost of most header runs in my day. It was built into the prices quoted for jobs such as our example. So, in this example, we would calculate the raw material cost as 945 pounds, times the purchase cost of the wire say $0.55 cents per pound, or $519.75. In actual fastener manufacturing costing, transportation and material handling costs might also be added; however, the material costs would be proportional. Similarly, the nut cost would be developed and used.

Although labor relations can be very difficult at times, the labor component of fastener manufacturing cost is relatively straight forward. We find the hourly wage of the cold header operator at xx dollars per hour. Next, we determine the number of cold-heading machines being operated by this operator, let us say three in this example, as well as the number of ordered parts produced per hour. We will use a header speed of 60 ppm with a production yield of 55 ppm to account for punch changes, coil changeover, and miscellaneous down time. Our final labor calculation would be an overhead or "burden rate," which factors in nondirect labor such as maintenance, tool room, quality assurance, shipping room, and administrative personnel. A typical burden rate might be 1.75 times direct labor. In year 2002 dollars, that may add up to a labor cost per hour of $70. At our planned production rate, we will plan to produce 1000 pieces every 18 h. Our heading labor cost is $70 per hour times 18 h per thousand parts, or $1272 per thousand, or $1.27 each in 1000 lots. Setup time would be included in production times. Fast machine changeover would be an economic advantage. Extended machine setup times would be an economic drag.

Threading labor would be similarly costed, as would heat treating and plating. That gives us our material and labor bolt costs. Nut costs would be accounted in a similar procedure. Finally, tooling is amortized over its useful life. If it costs $7500 to tool a cold header to produce this part, and the tools on average will produce 100,000 parts not counting perishable (or wear) tooling such as heading punches, then the heading tooling costs would be $75 per thousand, or $0.075 each. Threading tooling, usually less expensive, would be treated similarly. Adding these fixed and variable costs together, we obtain a manufacturing cost on which the fastener manufacturer can base a price with which they can compete for this business.

Looking at the consumer side of the fastener economic transaction, the fixed and variable costs are less well defined. Let us say that M25 bolts and nuts are to be used in a cell phone tower, which is to be sold to a public utility company when erected and operational. Each tower requires thousands of these fastened connections. We have 10 cell phone tower orders in our backlog. Let us address the variable costs first. The 10,000 fasteners will generally carry a purchase price that is based on the manufacturer's cost plus a reasonable profit margin to justify the use of capital. The manufacturer's idea of a reasonable profit may not be in accordance with the user's willingness to pay one cent more than possible to obtain the fasteners. There are usually enough suppliers of these fasteners that competition forces the price to an equilibrium price which gives a margin which justifies business continuance but not so generous that fastener factories are opening in every available industrial space in the town.

The variable cost also includes purchase orders, freight, stocking bins, and labor to move the fasteners into the proper locations in the tower assembly. Applying the torque and torque reaction to run the fasteners screw threads together and develop the specified clamp is a variable cost. A rule of thumb on these costs to purchase and install these fasteners is the 80/20 rule. The purchase cost of the fastener, say $2 each, is 20% of the installation cost of $8 each. For this reason, an assembler seeking to control fastener costs should apply effort to the cost variables in equivalent proportion. Fixed costs would include the wrenches, air or other power drivers, and wrench, gage, and measuring equipment calibration. Fixture, benches, shop air, installation drawings are all fixed costs associated with the assembly of these fasteners. Comparing the manufacturing and installation costs of this example, it can be seen that fastening cost can be a significant part of industrial commerce. Many innovative techniques such as stockless purchasing, consignment orders, quick-change tooling, lights-out manufacturing, and other industrial developments and innovations have helped reduce costs and improve fastener commerce to economies worldwide.

4.5 Engineering Economics

Engineering economics is the quantitative study of the money used and generated in industry. A useful model for the money flows in fastener commerce is the fluid flow model where dollars, or a local currency, are the working fluid. We measure this money flow with two parameters: dollars and time.

The first important principle in engineering economics is the time value of money. In an equation form, it is expressed as follows:

$$F = (1 + i)^n P$$

where
F = The future dollar value of an amount of money
i = The interest rate in effect for this transaction
n = The number of compounding periods
P = The present dollar value of the amount of money

The equation gives the future value of a sum of money based on a length of time and an interest rate. Since we are raising the value by an exponent, the relationship between present and future values is exponential. The relationship is dependent on the interest rate over time. Let us look at it graphically in Figure 4.8. Notice that for a positive interest rate, over a length of time, the monetary value will be higher in the future. For this rate of interest, the longer the time interval between present and future, the larger the increase in monetary value. This is a graphic representation of the time value of money. If we invest x dollars in tooling to produce a component, or we purchase a bond of fixed maturity, say 1 year at 4%, we would expect the bond at maturity to be 4% more dollars, and the tooling to cost 4% more if costs have increased by this amount. As the reader can clearly observe, interest rates and cost changes are rarely "sure things." A negative interest rate would

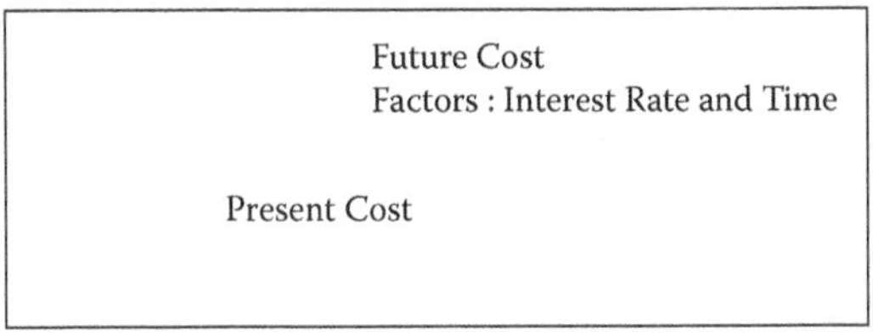

Future Cost
Factors : Interest Rate and Time

Present Cost

FIGURE 4.8
Future to present costs.

indicate a depreciation of future value. However, by the expected size of the interest rate, and the time period length, value changes. Rarely, if ever, do monetary values remain the same over time.

To illustrate this equation, we will use two interest rates, 4% and 6%, and two time periods, 4 and 6 years in the equation and graph. Our present value will be $1000. First, we will solve the Future Value equation for our four combinations:

1. $i = 4, n = 4$

$$F = (1 + 0.04)^4 \$1000$$

$$= \$1169.86$$

2. $i = 4, n = 6$

$$F = (1 + 0.04)^6 \$1000$$

$$= \$1265.32$$

3. $i = 6, n = 4$

$$F = (1 + 0.06)^4 \$1000$$

$$= \$1262.47$$

4. $i = 6, n = 6$

$$F = (1 + 0.06)^6 \$1000$$

$$= \$1418.52$$

Notice that in a comparison of the future values, both time and interest rate play important roles. Especially noteworthy is a comparison of the future values of examples #2 and #3. A lower interest rate for a longer time yields a higher future value than a higher rate held for a shorter period time period. Graphically, the four examples can be viewed in Figure 4.9.

Another engineering economics tool is the cash flow line. This can be seen in Figure 4.10. In this tool, monetary receipts are shown as upward vectors and cash expenses are shown as downward vectors. The larger the amount of money, the larger the vector. Distances along the line represent time periods. Let us consider an example. An investment is made at time zero and it yields dividends annually. The investment amount is $10,000 and the annual dividend is $500. Our time line would be like Figure 4.11. Combining these economic tools, the present value equation, and the cash flow line allows one analysis of a wide range of assembly operations. Let us consider two applications. One will be from a fastener manufacturer's viewpoint and the other will be from an assembly operations viewpoint.

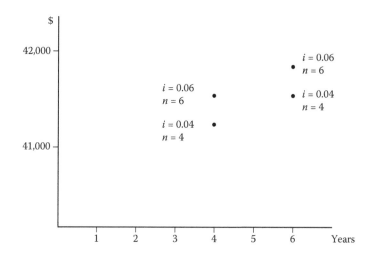

FIGURE 4.9
Cost plot.

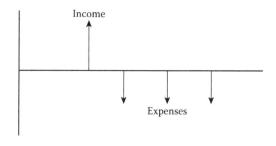

FIGURE 4.10
Fastener cost flow.

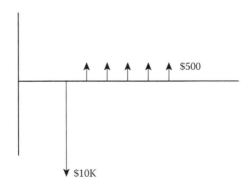

FIGURE 4.11
Fastener cost amounts.

Manufacturer: A screw and bolt manufacturing plant receives a request for quote for a hand tool component presently being hot forged. The sales engineer, responding to the request for quote sees that with the proper tooling and development, the part could be cold headed on one of the company's production cold headers. It will meet the part drawing dimensions with a simple machining operation of the cold headed blank. This can be performed in house as a secondary operation. The best estimate for cold-header tooling and development is $18,000. In preliminary negotiations with the customer, it is determined that the price which will convince the customer to convert from the hot forging will yield a profit of $2500 per order. Production estimates indicate an order for this part twice a year. The customer confirms that the job will run for at least 5 years. The company can earn 4% on treasury bills and similar fixed income instruments to provide a hurdle rate for uses of the funds and that rate internally as a rate of interest basis. Looking at the cash flow line, we see outflows and inflows as in Figure 4.12.

A calculation using $i = 4\%$, a present value of $18,000, and a value of n equaling 5, we see the calculation as follows:

$$FV = (1.05)^5 \$18,000$$

$$= \$21899.75$$

Since our accumulated profits are $25,000, this appears to be a good job to quote, provided our estimates are reliable. A bonus is that we will be receiving a steady stream of inflows as opposed to a lump sum 5 years out.

Assembly operation: A manufacturer of a consumer product has been using an assembly line with air-powered hand tools tightening screws by operators at assembly stations. Volume has grown to the point that four operators are working overtime to assemble product orders at peak demand. A

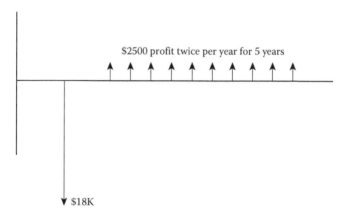

Compared with $18K and $400 per year with treasury bill at 4% interest

FIGURE 4.12
Relationship of cost and profit.

proposal has been received for an automatic machine which will produce 1.5 times the peak assembly output of the four operators. Each operator earns $24,000 per year without overtime. The machine costs $65,000 to purchase and install. It will require one operator to tend the machine. Other manufacturing duties will be performed at least half of the time by this operator. Three operators will be reassigned. The company uses 7% as its internal interest rate basis. Is installing the automatic assembly machine justified based on an engineering economic analysis? Let us look first at our cash flow line for this application as seen in Figure 4.13.

Performing a future value analysis, we see how compelling the economics are

$$P = \$65{,}000,\ i = 0.07,\ n = 1$$

$$FV = (1.07)^1 \$65{,}000$$

$$= \$69{,}550$$

Clearly, the machine will payback quickly, again based on the reliability of the information.

In summing up engineering economics for fastening, joining, and assembly, several key points should be emphasized. First, economic analysis is an important component in the decisions on how to fasten and manufacture. Also, we have just skimmed the surface of a very deep and complex area with these two simple applications. Cash flows are often more spread out and unevenly timed than presented here. Interest rates can also be more complex and dynamic. Finally, assumptions and estimates are just that. Reality usually can supersede

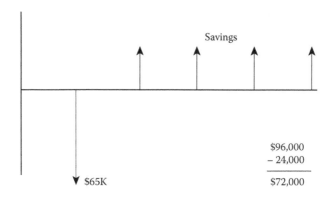

FIGURE 4.13
Cash flow line.

these with severe consequences for the unprepared and uninformed. In all cases, there is rarely a substitute in the long run for doing one's homework.

A note on business cycles while we are talking engineering economics is in order. I took a graduate class on finance. The students were, like me, working engineers. The professor made a statement that both bothered and intrigued me. He said: "A good finance person can generate more profit for a firm in one day, than an engineer can in a year."

A look at engineering economics would be well served to place it into the context of a fastener business's economic climate. In my experience, the business cycle is an important part of the fastener industry as well as of the businesses it affects as suppliers and customers. It will also have a somewhat significant impact on an individual's personal economic prosperity. The business cycle is the rise, plateau, and decline of economic activity. Some commonly used statistics are an economy's gross national product, or GDP. This statistics provides an estimate of the total output of the economy's goods and services. This includes the fasteners. When I wrote the first edition, the estimate of the domestic fastener business was $7 billion, which a small fraction of a very large economy. With this current edition, the nation's GDP is much larger, as is I suspect the fastener products and services. However, it has not been a steady, straight-line growth. It has gone up, and declined with sometimes gut-wrenching suddenness. Earlier, I introduced the extension bar application. In the 18–24-month period from the initial customer contact until the height of the production orders, the standard socket screw business went from very robust to very weak. Despite the best efforts of the sales force, incoming orders for screws and related catalog fasteners plummeted. Since that early experience, I have experienced it many times. I have lived and worked through many business cycles. They all were unique in time and characteristics. Most were understood most fully only in hindsight. Economics is usually divided into micro- and macroeconomics.

Macroeconomics is the economics of the firm. That can be the business you work in. Macroeconomics is the economics of all the economic activity— macroeconomics affects microeconomics. Many successful fastener engineers make it a point to have knowledge of economic trends and statistics. It can be used to change the approach to special projects such as extension bars, new fastener product development, and mundane economics factors such as which hurdle rate to use in pricing. The understanding that business cycle downturns are a fact of fastener life points to a few key areas. One area is being prepared. Those fastener firms which survive the downturn best, with the fewest layoffs and reduction in force, are financially fit. They also have a product portfolio of current good profit producing fastener lines and customers, as well as new one in development. They also have a watchful eye on those in decline. The old saying, "a successful manager knows when to kill a dying product, the unsuccessful manager dies with it," holds especially true in downturns. And above it all, the successful have both a firm grip on their core technologies and an equally keen motivation on serving their fastener customers very well.

I used to think and say, "We don't have to be perfect, just better than our nearest competitor." I still do, only more so as the survivor of many business cycles. Economics is important. Cash flow is the life blood of any fastener business.

Suggested Reading

1. Editorial, Tom Dreher, Fastener Technology International, February/March 1995, Stow, OH.

5

Assemblies under Dynamic Loading

When looking at the work of designing effective, efficient fastening systems and assembly procedures, perhaps no application or service presents more of a challenge than those subject to dynamic loading. These loads can be periodic or continuously changing in frequency and amplitude. The engineering mechanics are much less straightforward in their analysis.

In Chapter 3, we outlined some of the basic equations quantifying tension fasteners under relatively stable loading conditions. In this chapter, we will look at the reaction of fasteners to nonstable loads and offer some procedures for handling these fastening applications.

If we categorize some of the types of dynamic loads found in assemblies, we could list impact, alternating loads resulting in possible fatigue, alternating loads resulting in vibrational loosening, and thermals. As we examined in Chapter 3, in rigid metal joints, a well-established tension fastener design procedure is to apply a preload in the fastener, whether by the application of an adequate level of torque through an efficient thread and bearing surface friction, to elongate the fastener(s) grip to a preload higher than the service load, or by direct elongation of the bolt or screw, and compression of the assembly's clamp-affected zone (CAZ) to achieve this preload. If, instead of using vectors, we use a Cartesian coordinates graph, the fastener and service loads would look like those illustrated in Figure 5.1.

While this presents a tidy fastening solution, it does not address the real-world service load requirements of some alternatingly loaded assemblies. This may be due to the natural flexure of the assembled components, or the nature of the service loads in these applications. In this case, the loads could look like those shown in Figure 5.2.

An application in which service loads rise sharply, peak, and then subside would have a service load curve which looks like an inverted V. This type of load could be typical of many manufacturing processes. A stamping press, for instance, would see a sharp rise in service load as a work piece is struck by the tooling during the forming blow of the tooling. The load rise and peak may only occur for a slight fraction of a second and a few degrees of press crankshaft rotation, in the complete cycling of the stamping press. Another similar application the author has experience in also performs a stamping manufacturing operation, but in this case it presses out silicon wafers, part of which is accomplished with pins made from a specialty bearing steel to withstand the high, impulsive loads applied. The service load would look like Figure 5.3.

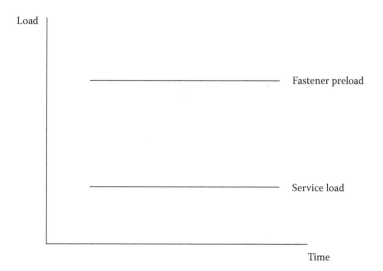

FIGURE 5.1
Service load and preload in well-fastened rigid joint.

Another load profile that can be found in some dynamically loaded joints results in a load that increases in load amplitude steadily with time. There can be several application factors that can result in this type of load pattern. One is the additional load placed by wear in the application components. For example, wear in bearing members and supports as a product's service

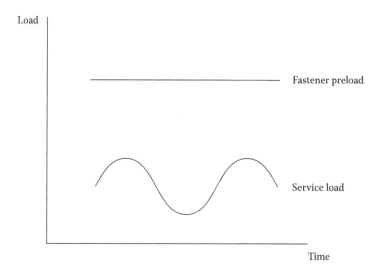

FIGURE 5.2
Alternating service load.

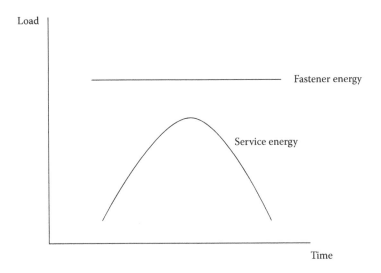

FIGURE 5.3
Impulsive loading.

hours increase can cause the load to rise. Another is the load increases resulting from component geometry changes due to temperature changes. A case typical of this would be one in which a temperature rise, $A\Delta T$, results in a load increase, ΔL. Figure 5.4 represents this type of fastener service loading.

A final dynamic fastening load we can consider increases with time as the previous service load function, but at a greatly increased rate. This particular

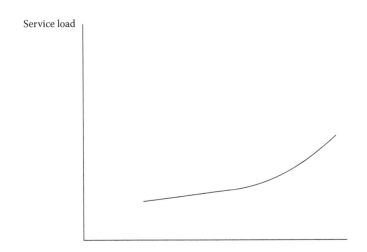

FIGURE 5.4
Time increasing load.

Service load

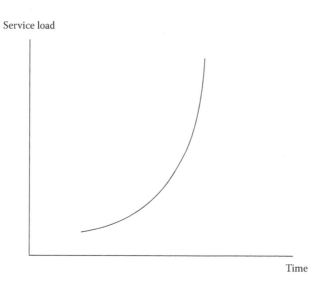

Time

FIGURE 5.5
Rapidly increasing loads.

loading pattern may not be desirable, but may nevertheless require fastener provision for it. Some examples would be in redundant systems as a fail-safe or backup fastening system for a companion fastening system. A schematic sketch is shown in Figure 5.5.

Applications where we might expect to find service loads of this load profile would be in areas where vibrational loosening is a possibility or where a component failure could suddenly shift a rapidly increasing load to a different or redundant backup system. One set of application that comes to mind is the automotive hood and door latch assembly.

To summarize the different types of dynamic loading fastening systems can be reasonably expected to handle, along with some classic fastening examples of these applications, we can list the following:

Sinusoidally alternating	Engine connecting rods
Short-duration peak loading	Stamping punch press
Low increasing rate loading	Mechanisms with time-related wear
High increasing rate loading	Temperature-related assembly member dimension change

We can examine the fastening approaches that can be taken to minimize the effects of these dynamic assembly loads and illustrate some fastening processes and products, which can be effective in dynamically loaded assemblies. A good starting point is to consider the role that the fastener and assembly materials play in absorbing and dampening dynamic fastener service loads.

5.1 Mass Effects

The material characteristics, such as density, and the fastening and site characteristics, such as geometry, play a vital role in the response of fastener-joined assemblies to dynamic service loads. There are two aspects to the nature of response. One is the damping characteristics of the material and its geometry about the site. By dampening dynamically applied service loads, the energy from these loads is absorbed and its effect dissipated. The molecular structure of the material is a cushion. In a simple two-plate and screw assembly, if the plate and screw materials have high-energy-absorbing characteristics, more dynamic energy in the form of changing load carrying can be an assembly benefit.

The second aspect of the mass effects of fastener materials is the inertia of these materials. The same characteristics giving assembly benefit in the form of dampening and energy absorption may bring with them a high inertia. This inertia can be detrimental in the case of fastened assemblies, which we require to accelerate and decelerate as part of their operation. They also can be detrimental in more commercial ways with regard to procurement, assembly, and shipment. Clearly, a fastener material that has infinite energy absorption coupled with zero weight would find a ready set of assembly applications. Unfortunately, one still cannot get something for nothing and "Unobtainium" has yet to be developed.

Fortunately, when designing fasteners and systems for the dynamic load curves shown earlier in this chapter, steps can be taken to achieve more than satisfactory fastening performance.

Let us look at some good fastening design examples of using the mass effects of fastener and assembly materials to advantage in dynamic service loading applications.

5.1.1 Four-Bolt I.C. Engine Main Bearing Supports

Many automotive engine crank shafts run in main bearings that are clamped together with a cast iron cap and two main bearing cap screws. In some cases, pins are used in addition to provide location and minimize side loading. In themselves, the screws provide the clamping power but it is the sum of the bearing cap mass plus the screws' holding power and additional mass, which absorbs the dynamic stresses, in this case cyclically applied. This is a key concept in the design of fastening systems for dynamic loading. It is the combined mass of the unit's components, sometimes strengthened with additional mass geometry, and the fasteners, which provide the "mass" reservoir to handle the loads as they vary during functioning.

Returning to our engine example, during the "horsepower wars" which occurred during the late 1960s, the major automotive manufacturers were greatly increasing the output of their engines. This in turn greatly increased the stresses on the engines' internal assemblies. This included the crankshaft main bearing caps and cap screws. One innovative, and very durable, design of the time was a "four-bolt" main bearing, which as the name implies used four fasteners rather than the conventional two. An additional two cap screws were torqued into tapped holes in the crank supports at right angles to a more normally positioned two screws. This "boxing" effect proved quite an increase in strength over the conventional configuration, in no small part due to the additional mass that was brought into action around the highly stressed bearings. Bringing the mass effects concept up to more current designs, the use of designed geometry machine tool bases accomplishes an increase in dampening while at the same time optimizing the amount of raw materials used.

5.1.2 Mass Effects of Oil Pan and Gasket Screws

Another engine example that demonstrates the dynamic energy-absorbing effects of an assembled mass are the threaded fasteners that hold the oil pan and gasket to the underside of an engine block. Vibrational energy can present a challenge as can the inclusion of the sealing gasket taking away some of the rigid joint characteristics of the joint sites. By extending the block casting down to below the crankshaft centerline, additional casting mass is brought into action. This additional mass dampens and attenuates some of the forces emanating from the cylinder explosions and bearing reactions. This reduces the vibrational forces exerted on the pan, gasket, and pan fasteners. The additional mass adds to the stability of the fastening system.

Of course, this additional mass comes at a cost and cannot be used indiscriminately. Cost and competitive pressures in addition to a need for economy and efficiency preclude it. However, modern techniques and innovative methods can place needed mass in an assembly without resorting to excess.

One recent example of a well-designed, dynamically loaded joint comes from a European automotive connecting rod. The connecting rod is cast in one piece but unlike conventional practice, the rod is "snapped" apart so that a jagged surface exists between the rod sections. Since the surfaces were one prior to snapping, they fit together, crevice for crevice. The sections are screwed together after snapping to provide a clamped joint surface which, while no larger in area than a conventional rod, brings considerably more rod mass into action under dynamic loading. This is a good example of mass effects and truly innovative fastening! Certainly in the area of engineered molded plastic component assembly, the judicious placement of polymer around a fastening site can increase the dynamic strength and load carrying of the tightened joint.

Consider the design of two injection-molded plastic screw sites. Design 1 has a wall thickness around where the screw will thread into, which is at least two times the hole diameter. There are no supports around the site. Dynamic loading starts when the screw is turned at speed, perhaps 30 rpm or higher, into this site. With such minimal mass, overloading of the polymer is a consideration. Now consider design 2. Wall thickness is at 3× hole diameter, and stiffening ribs have been added. The addition of mass in design 2 provides for more dynamic load absorption. These can provide the assembly and its consumer with additional robustness and utility in the assembly's functioning under real-world dynamic loading conditions.

It is important when designing and specifying fastenings to consider the entire affected zone, or volume, of the assembly. There sometimes exists the tendency, often with existing designs that are being reworked, but sometimes also with "clean paper" designs, to want one special "quick fix" special fastener that will solve all dynamic loading problems. My contention is that this is not the best approach. Even if a fastener is developed and sourced for the application, it is not making efficient use of the assembly or the assembling company's resources.

A better approach might be to take a systems view of the entire assembly, especially that area around the fastening site and use its mass to perform fastening work. By taking first a systems/mass design view and then carrying it out in assembly, more dynamic load can be carried with less or more efficient fasteners.

Returning to our automotive application. Let us say we are designing a fastening system to assemble the engine subassembly to the main car body/chassis. One inefficient approach would be to simply design brackets to the engine block sides and bolt these to the tops of chassis frame rails. The bolts, brackets, and frame rails will carry the load, provided we make them of dimensions and mass material conditions, namely strength, ductility, and endurance limits, to absorb the service life loads. Let us now consider using more than just the narrow area of the area of the chassis on which the engine brackets bear. We could make a V-shaped cradle of the section of the chassis where the engine will mount. This would form the two equal-length legs of an isosceles triangle. By shaping the engine brackets with matching angles, we can make the mass of the engine block a stressed member of the assembly. The fasteners we use will hold the bracket joint faces together but will be greatly supported by the much increased mass of engine block and chassis area in dynamic load carrying.

Using this mass effects fastening approach, we can transfer this approach to business machine motor mountings, vending machine components, or any of an unlimited range of dynamically stressed parts. The ingenious fastenings section in Appendix B shows some examples. Before looking at some of the fastening and testing equipment, techniques and procedures for dynamically loaded assemblies, it is worth looking at the area of stress risers in fasteners and components under dynamic loading.

A stress riser is a notch, crack, scratch, or dimensional change that causes an applied service load to be amplified in the area of the stress riser. They are of interest and importance to the fastener designer and user in that under the load changes accompanying dynamic loading, these raised stresses can bring about failure from fatigue. A simplified example can illustrate this phenomenon.

A paper clip is a very simple fastener. It is a metal, has large smooth radii, and has very low stress risers coupled with a high design factor (strength-to-load ratio). If we test the clip in tension, perhaps by testing one of the longer straight sections, we will determine the yield and tensile strength along with perhaps the elongation and reduction in area.

But what if we planned to place this "fastener" in a hypothetical dynamically loaded application. We would be more interested in the clip's strength relative to the dynamic loading profile, be it fatigue, impact, slow or fast rising rate, or thermal. Thinking the service load to be one factor that might tend to precipitate a fatigue failure if not designed and fastener for, we would want to test the clip to determine its fatigue capabilities. Certainly, it can be pictured that bending the clip repeatedly will ultimately cause it to fail from fatigue, accompanied by a rise in temperature in the bending zone caused by the pin's mass-absorbing load energy. If we formalize our testing, we can document and replicate the degree, amplitude, and direction of our bends, their frequency, and record the number of cycles to failure.

If we introduce some stress risers into our clip fastener, say by placing a large scratch, notch or break in the bending zone, we intuitively know that the number of cycles to failure drops significantly. Our endurance limit and fatigue strength have decreased markedly.

This is the concept of stress risers in dynamically loaded fastener systems. Notches, marks, or sharp geometries of all types raise the service stresses at these locations and can greatly diminish dynamic strength and usable service life. It is important in applications where service loads fluctuate to control the stress-raising features of the fasteners used. Several good steps include generous fillet radii, reduction of tool marks and manufacturing and assembly produced scratches, nicks, and gouges. Figure 5.6 shows some of the stress risers possible in a hex head machine screw.

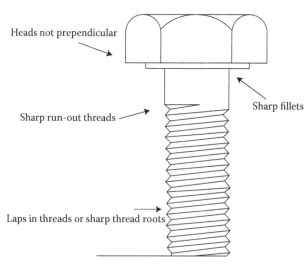

FIGURE 5.6
Hex head screw with stress risers indicated.

As can be seen, these and similar stress risers in other fastener types can be detrimental to the dynamic load application performance of a fastening system. These small features can sometimes be the source of crack propagation and assembly failure.

Users of fasteners for dynamic loading can take several procurement steps to upgrade the product integrity of their assemblies. First, test to the greatest extent possible using the fastening installed in the assembly. The remainder of this chapter will review dynamic fastening testing. By using testing of your own configuration, or the services of a third-party laboratory, you will have taken important steps toward increased performance, and you will know your sources and their fastener's quality.

5.1.3 Testing for Dynamic Loading Applications

Where dynamic loading is a known service condition, testing is usually fundamental to the fastening design process. With dynamic testing, the lab or test site conditions should simulate the real service environment to the greatest extent possible. As a way of demonstrating the necessity of application of real duplication of fastening dynamic conditions during testing, the window attachments in a famous northeastern city office building offer good instruction.

This high-rise commercial building started to experience window glass being pulled by high wind loads from their fastenings to the building structure. Several fixes were attempted and the building subjected to much engineering analysis. It was only after a researcher went back to the wind tunnel model and added models of all of the surrounding buildings and structures that the true amplitude of the wind loading showed up. The surroundings created vortices that accelerated the wind past the windows. The dynamic loads were higher than the original tests had predicted and the window attachments had been designed to fasten against. Based on this new data, reinforced window attachments from the new dynamic loads were successful.

Before we examine some dynamic testing equipment and procedures, which may prove of use in your assembly designs, a more recent example of dynamic testing can illustrate the importance of recognizing dynamic service loads and testing fastener systems and assemblies for their ability to successfully withstand them.

A large consumer products company was designing a new model of one of their most successful products. The injection-molded main assembly body used some of the most advanced injection mold-making and injection techniques available anywhere. To minimize parts count and control assembly costs, all components were designed for snap fit to the main body. No other fasteners were used. This would also take advantage of the close tolerances made possible by the mold-making techniques and equipment. The product did snap fit together as planned and appeared ready for a successful launch.

This company prided itself on its reputation, which was well earned, for customer satisfaction and product value. Before going ahead, they devised a simple but effective table to simulate the shaking and handling that the product would see during its useful life. The snaps loosened. Dynamic testing prevented a fastening problem.

5.2 Vibration

Many assemblies experience vibrational loading as a function of their normal operation. And yet testing for fastener vibrations is not a common event outside of fastener and commercial laboratories. In shipbuilding, one of the final phases is the "shakedown" cruise which determines how systems work out at sea. Some automobile companies constructed proving grounds with cobblestone or Belgian block sections which vibrated the entire auto as it drove over them.

One well-established vibration tester used in the fastener industry is known as the Junkers Transverse Vibration Tester. It is named for the German engineer who developed it in conjunction with a leading socket-fastening products company. It was well used to test out the vibrational resistance of the company's self-locking fastener thread forms under varying conditions of preload and vibration, among other variables.

The Junkers Transverse Vibration tester is driven by a fractional horsepower electric motor. The motor is connected through a Pittman arm to a steel block which is free to move in a horizontal plane between limits set by the test technician. The drive block is confined vertically by a set of "ways" which guide its reciprocating travel. The fastener being tested is installed with its axis perpendicular to the line of travel of the tester loading block, hence the name, transverse tester. If we construct a simple diagram of the test conditions we would see one similar to Figure 5.7.

The amplitude of the vibration force can be varied over a range of tests to determine the vibration threshold that starts to unload and loosen a fastener. It is important to remember with any threaded fastener that the threads are inclined planes wrapped around a cylinder. A block resting on an inclined plane, held by gravity and friction, will start to slide if interrupted by a banging force on the end of the inclined plane. Similarly, a bolt or screw, preloaded and holding fast, can be "walked" out of its mating thread by a banging force somewhat on a plane with the threads. A typical machine screw thread helix angle is 4° and some number of minutes. The Junkers tester brings the vibrational energy close to the thread plane.

As a vibration tester, it is effective in that it enables a fastening system and its vibration resisting elements, such as locking threads, chemical thread locking compound or mechanical locking feature, to measure specific

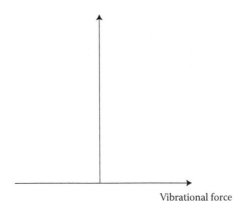

Vibrational force

FIGURE 5.7
Fastener clamp and vibration loads.

levels of force to determine capability under given application conditions. It also has provisions for finding the vibrational thresholds above which unfastening occurs. Figure 5.8 shows a plan for a Junkers-type vibration tester.

If your assemblies are going to be experiencing vibrational loosening forces and you are either in the process of specifying a vibration-resistant fastening product or need to perform a comparison of locking methods, a Junkers transverse tester makes the process of differentiating vibrational resistance easier to evaluate and quantify. Whereas it is not a test covered by a well-circulated industry standard, it is well established. Another method that can help subject a fastener type or fastened assembly to vibrational testing is to induce dilation of the mating threads by striking a force

Fastener to be tested

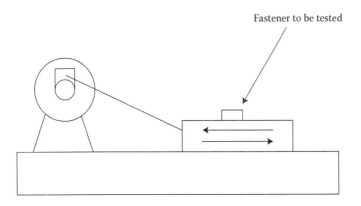

FIGURE 5.8
Fastener vibration tester.

on the end of a protruding threaded end of a fastener. Good engineering practice calls for at least one full thread to be visible beyond the nut face farthest from the head of the screw. The bottom of these threads provides a good striking area provided it is delivered with an anvil, which transmits good force without damaging the threads and preventing thread disassembly after the test. Nut dilation is a hoop stress phenomenon. In metal fasteners and components, the axially applied forces generate a secondary Poisson's ratio force, which radiates out first through the external and then the internal thread.

In a vibrational test setup using dilation, an air-powered cylinder or similar device is arranged so that it delivers its force with an amplitude and at a frequency established to simulate the vibrational forces developed during actual service operation. Figure 5.9 indicates the general setup. A similar device is within the capabilities of most shops.

One final vibrational tester that can serve as a quick vibrational tester is the familiar "paint mixer" type, similar to the kind found in every paint retailer for agitating the contents of paint cans. A simple mechanism of this type can be adapted to serve your dynamic testing.

Fastener being tested

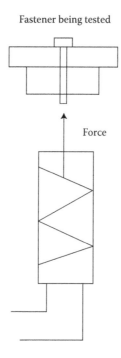

Force

FIGURE 5.9
Fastener vibrational/dilation test setup.

5.3 Fatigue

Fatigue testing is a well-established testing procedure documented in both ASTM and MIL-STD testing documents. Fatigue is failure of a fastener or assembly under the action of repeated stresses. Fatigue testers can be found in many commercial laboratories and academic institutions. In a fatigue test, the fastener or assembly is normally preloaded to a predetermined mean load and the test load is alternated about this mean. Normally the test load is sinusoidal, with the load first going above and further straining the test material, and then unloading and unloading both the incremental test strain plus some of the preload as well.

In one 5/8 inch connecting rod screw fatigue test requirement the author is familiar with, the mean load is 30,500 pounds. The print specification requires an alternating load of ±1500 pounds with no failures after 10 million cycles. As a qualifying procedure, new suppliers were required to demonstrate on the fatigue tester an endurance limit of ±3000 pounds, or twice the specification. With attention to stress riser control, thread quality, finish thickness, and some tweaks that will remain confidential, the required endurance limit could be reached by any competent bolting manufacturer.

The test data from a fatigue test gives the endurance limit, or the number of cycles the test specimen lasted at the given maximum stress before fatigue failure. A set of coordinates are plotted on an S–N diagram. Semilog paper is used with the number of cycles being logarithmic. Ten million cycles is considered infinite life. If a test specimen lasts for 10 million cycles, it is considered that fatigue failure is unlikely.

Two types of fatigue testers are the rotating beam and the resonant frequency. In a rotating beam fatigue tester, the loading arm rotates at a specific speed, loading and unloading the specimen once for each rotation. For a typical tester, a rotation rate of 1800 rpm would be common. At this rate, 10 million cycles would require just under 4 days. The machines sometimes have the capacity to test more than one specimen at a time. If several test coordinates were to be plotted for an S–N plot, this would be a time-consuming approach.

The resonant frequency tester compresses the time required by determining the mass and resonant frequency of the specimen area and then varying the fatigue-causing forces until the specimen is in resonance. Circuitry then converts the resonant fatigue failure to its rotating analogy. Figure 5.10 shows an S–N curve. This is one area of fastener/assembly design and testing where modern microelectronics has made a significant contribution to the dynamic testing and performance of highly stressed fastener assemblies. It is often a combination of modern technology with tried and true mechanical engineering that increases assembly performance.

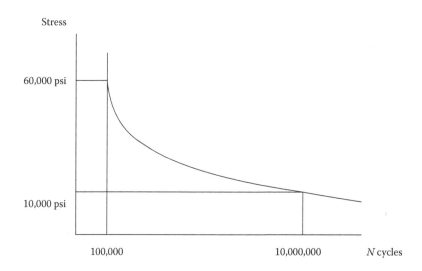

FIGURE 5.10
Fastener fatigue *S–N* curve.

The fatigue tester can also serve a useful function to the metallurgist who examines the fatigue fracture planes of the test samples. The location of the fatigue fractures on the fastener, their mode, and rate of propagation can help give a strong indication of areas of possible improvement in the fastener manufacturing process. For example, if fatigue test failures occur consistently in the thread areas, which is a common design occurrence, manufacturers can develop thread rolling tooling and processes to smooth and lower stress risers in these areas, or roll after heat treatment for offsetting compressive stresses.

5.4 Some Additional Dynamic Tests

In addition to fatigue testing, several other dynamic load-testing techniques and equipment are available if the application will have service loads, which they simulate. One of these is the impact tester. Figure 5.11 shows an impact tester. In fastening service where extremely low temperatures are to be encountered, impact strength of the fastener and fastened materials is of critical importance to assembly integrity.

A notched specimen of the material to be tested is placed in the impact tester and a mass on a pendulum is released to accelerate through an arc and strike the test specimen. There are two types of impact test, the Charpy and the Izod. In each, the energy absorbed in fracturing the test piece is measured. The impact velocity in the Izod test is 11.5 feet per second (fps). An impact velocity of 17.5 fps is used in the Charpy impact test.

FIGURE 5.11
Impact tester.

Since no direct fastened impact data is provided by the standard Charpy or Izod test, their value is in the information they provide the assembler, designer, and manufacturer about the performance of specific fastener and assembled materials under the dynamic loads of impact. These can then be used to make assembly materials decisions.

Variations on the standard impact test are two application-specific tests that have been used in consumer product evaluations. The first is the crash test, familiar to everyone as the test of automobile bumper systems. In the testing of an assembled product, some mass, say a steel ball, is dropped from a standardized height, for example, 3 feet, onto a specific zone on the fastened assembly. The assembly designer can assign a drop test success criteria. One such criterion would be the ability to continue proper product function after one drop application.

Another test application would be one that simulates the unfortunate accidental impact that confronts some assemblies is the unintended free fall. Many products are inadvertently dropped and it is good application information to know the condition of a fastened assembly after this type of dynamic loading. The important test factor in making comparative fastener evaluations is to standardize the test procedure so that to the extent possible, only the assembly variables are evaluated under the dynamic conditions associated with the drop. Some suggestions would be for the assembly to be released from a release height of 1 m and allowed to impact onto a safety cage enclosed standardized test surface of either concrete or steel. By standardizing the test procedures, test-to-test results will have engineering validity. Earlier, the test reference was made to lower temperature extremes for a tested assembly. If service loading will be accompanied by the dynamic factor of elevated temperatures, a creep-type test may be an important dynamic loading test to perform on proposed assemblies for this service.

A creep test consists of a weight applied to the test piece through a system of levers. The specimen is in a furnace and held at a constant temperature. The length-wise deformation is measured throughout the test. A curve can be plotted after sufficient test conditions to determine the response of a fastener with time at a specific service temperature. The loss of fastener strength and clamping preload can then be used as a design factor and allowance can be provided for long-term creep. Other fastener tests which can be conducted are stress-rupture, stress-relaxation, and constant strain rate.

Before concluding dynamic fastener testing, other tests that should be covered include the stress durability test for embrittlement, the wedge tensile test, and fastener drive and bend tests.

The stress durability test consists of loading a sample of a tension fastener lot to a specified, high preload under controlled conditions and leaving them at load for 24–200 h depending on the version of the test. At the tests conclusion, the fasteners are unloaded and examined under magnification for indications of stress cracking. Fasteners that have become embrittled from the concentration of atomic hydrogen in cleaning or electroplating will not usually fail during the initial tightening but will have a delayed, brittle failure. The stress durability test screens for a hydrogen embrittlement condition and is a useful test if fastener materials or processes have a possibility of hydrogen embrittlement.

A wedge tensile test is similar to a standard ASTM tensile test; however, a 6–10° wedge-shaped washer is placed under the head of the fastener. The wedge washer has to be well made with no sharp edges to cause additional stress risers in the tested fastener. The advantage of the wedge tensile test over the conventional test is that it tests the quality of the head to shank fillet and the fastener material's ductility, which can be important predictors of fastener and assembly dynamic load response.

Fastener bend tests also place bending loads on a test specimen held in a vise as a quick measure of ductility. Drive tests simulate the installation of fasteners with measurement of torque, rotation, and loads for a quantitative assessment of actual assembly performance.

5.4.1 Some Thoughts on Dynamic Loading

Dynamic loads are present in a great many fastened assemblies. It is only in the degree of dynamic nature that the loads should be considered to require special design attention in fastening. As a rule, the assembly producer should know the range of loading that is conceivable, design the fastening system to that high standard, and test rigorously and without bias to verify fastening integrity under dynamic conditions in the field.

5.5 Difference between Static and Dynamic Strength

It is human nature to simplify ideas. This makes them easy to understand and communicate. In engineering, this is useful to a point in that a complex concept, such as thermodynamic entropy, can be grasped at a very basic level, by a wide population. This is a good thing. The risk, however, is that oversimplification can lead to the concept being misunderstood. Such is the case in metal fasteners and the concept of mechanical strength. Tensile strength, which is a static property, is often misunderstood by fastener users, distributors, and engineering personnel. Dynamic strength is an important concept for an understanding of fastening. To best develop fully a more complete knowledge of the strengths of metal fasteners, we will study a graph plotting a piece of low-carbon steel loaded in pure tension. We will then examine a plain paper clip which, as we all know, is a fastener for holding papers together.

Our material will be C-1006 low-carbon steel. We will remember that steel is an iron (Fe)-based metal with carbon (C) added. The addition of carbon into the iron, along with manganese (Mn) and some other elements, binds the molecular structure of the steel so that it is stronger than pure iron. Let us look at a stress–strain diagram for pure iron, such as a blacksmith would have at one time made into a nail or other wrought piece of iron hardware. Let us also look at some commercial grade C-1006 available from a steel supplier in the form of bar or rod form. These can be seen in Figure 5.12.

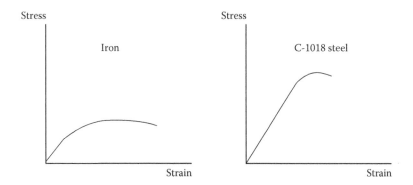

FIGURE 5.12
Stress–strain curves.

The stress–strain diagram plots deformation versus stress. For every unit of stress, we place on the iron and low-carbon steel coupons, they deform. At stresses below each material's yield strength, this deformation is primarily elastic in that once the forces causing the stress are removed, the coupon returns to its original shape and dimensions. If our test pieces have geometry changes, such as a machine screw thread, local yielding will occur at the more highly stressed areas at a general load below overall yield. With respect to dynamic fastening strength, the geometry of each plot is of interest to us. Notice how in each case, stresses and strains increase at the modulus of elasticity, or rate of deformation, up to the material's general yield point. Further note how each then starts a much more gradual increase until the curve's apex of ultimate tensile strength is reached before finally declining to the breaking strength. Between yield and breaking, the test coupons' areas are decreasing. On most tensile tests, a standard loading speed such as 1.00 inch per minute of cross head speed is established as a standard. Since area is a factor dependent on the diameter raised to the second power, this "necking down" before breaking is exponential. We can observe that the two curves have markedly different shapes after yield. This is a key concept in understanding the dynamic strength of assemblies and their fastenings. If we just consider tensile strength, we look at the height of the apex against the graph's vertical axis and make a comparative judgment. But it is in the area under the curve where the ability to absorb dynamic energy and maintain holding forces that dynamic fastening ability should be evaluated. This is a two-dimensional measure since we are considering both the vertical axis, how much stress did it hold, and also, how much was it able to deform before breaking? Now let us take two lots of C-1006, with differing areas under their stress–strain curves and fabricate some paper clips. The stress curves are seen in Figure 5.12.

Notice how Lot B has a higher ultimate tensile strength. If we were asked which steel will make the stronger paper clip, ceterus parabus (or all other things being equal, in Latin), how would we answer? The answer is Lot B if the tensile strength is the only criteria. However, it should be Lot A if we were going to use metal fatigue as the test of strength. We are all familiar with the common paper clip shown in Figure 5.13, however, let us evaluate it in more technical detail.

Notice that the diameter of the wire from which it is manufactured is 0.031 in. The calculation shows that its circular area is 0.00075 in^2. Based on yield and tensile strengths of lots A and B, our paper clip yield and tensile strengths are shown in the figure. Now let us look at our test apparatus in Figure 5.14.

FIGURE 5.13
Paper clip as fatigue strength test specimen.

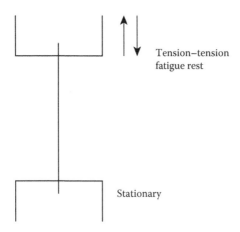

Tension–tension
fatigue rest

Stationary

FIGURE 5.14
Fatigue test set-up.

The tensile tester applies a straight pull perpendicular to the wire circular area. This would be an unusual method to break a paper clip. The fatigue tester breaks our paper clip in a method most can readily see if not fully understand. By bending the paper clip repeatedly, a force smaller than that required to cause a tensile failure will result in fracture around the bend area of the wire. If instead of bending, we pull the wire repeatedly, this will also cause a fatigue fracture at some combination of pulling forces and cycles. These two mechanical tests illustrate the difference between static and dynamic strength. The tensile test type of loading, a one-time pull to failure, is less frequently found in tensile, compression, and shear-jointed assemblies than in fatigue-type loading. Returning to our question of lots A and B of raw material, we can see that higher tensile strength means a stronger tensile test result. But as tensile strength increases, ductility decreases. Viewing the stress–strain diagram, ductility is the area under the curve. Steel, or most component materials, have an optimum range within which both tensile strength and ductility are at optimally high levels. Lot A, having more ductility, will be able to withstand more loading cycles before failure in fatigue. Since resistance to failure from fatigue, as well as impact strength are important fastener dynamic properties, it is important to know not just the static tensile strength of a fastener, but also the dynamic ductility of a component's material. While there is not a ductility scale for steel and other fastener metals, elongation and reduction of area are good measures. As an example, for a 180,000 psi tensile strength alloy steel, it would have an average Rockwell "C" scale hardness after heat treatment of HRc 40, but more importantly for fatigue strength, it would have an elongation of around 12% and a reduction in area of over 30%. If our lower carbon steel could be brought up to a tensile strength of 180 KSI with a combination of cold working and case hardening, we would find it so brittle that its lack of ductility would make its fatigue

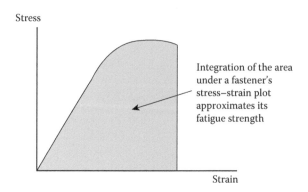

FIGURE 5.15
Area under the stress–strain plot.

strength unacceptable for any practical use. If we made it with a minimum of stress rising features such as screw thread roots and transition threads, its endurance limit could be expected to allow it to withstand an alternating tension–tension load of 10,000 pounds for 10 million S-shaped cycles. As a final observation, integrating the stress–strain plot as shown in Figure 5.15 would approximate the area under the stress–strain curve for a fastening component and help indicate its fatigue strength and useful service life.

The fracture plain resembles a pebbled surface. A good martensite grain structure will exhibit a surface such as this when loaded past its yield strength and then loaded to fracture. The necking down from yield will be evident. If this is in the threaded area, the lead of the thread will be elongated. A quick comparison with a good thread of the same diameter and pitch will show the yielding.

By comparison, the fatigue failure fracture plane will show no such gross yielding. The fracture plane may exhibit characteristic "beach" marks somewhat reminiscent with the marks an ebb tide leaves in the sand of a beach. Depending on ductility of the raw material, the stress risers in the component, and the loading cycles in service, these beach marks can be evident in the fracture planes of failed fastener components across a percentage of the entire surface area until it becomes so small that it fails in yield tensile. For this reason, always protect any metallurgical surfaces from failed components. And understand that fracture prevention is dependent on a set of interdependent factors of design, manufacture, installation, and service.

5.6 Fastener Chemistry

An understanding of chemistry plays an important role in the understanding of fastener application engineering. A good starting point is the periodic table, with fastener-related elements such as iron (Fe), carbon (C), manganese (Mn), and all of the other known elements arranged in a table form. The columns group elements into groups with similar properties such as those existing primarily in gaseous state or having metallic properties. Each element has an atomic weight and an atomic number. Chemistry and chemical processes are factors in the fastener engineering in raw materials and fastener manufacturing processes such as heat treating. Chemical processes are also at work in fastener applications in areas such as hydrogen embrittlement, and the corrosion of fasteners and fastened assemblies. Atoms are composed of a nucleus holding protons and neutrons and are orbited by electrons. The electrons have negative charges and the protons have positive charges. When operating as a balanced atom, the charges counteract each other. As the electrons are in orbital motion, they have centrifugal forces from the dynamics of this motion. This centrifugal force has a reaction in the centripetal force of the protons in the nucleus. Fastener chemistry is primarily about how the elements found on the periodic table react with each other in fastener manufacturing and application. Some example of fastener chemistry in action is corrosion. Substances can be solids, liquids, or gases. We tend not to think of the air all around us as a gas but it is. Air is a high percentage of nitrogen as well as significant amounts of oxygen and other "trace" elements. Steel is solid at room temperature and composed of a lot of iron, a small amount of carbon, and other elements such as manganese. Putting the iron in steel in a conducive chemical environment, we get Fe_2O_3, or rust. This is a chemical process. Writing chemical equations and balancing the weights on both sides of the equation is a great practice exercise to understand the chemistry in your fastener applications. Aluminum does not rust as steel does but it does corrode. The corrosion product is a white powdery substance. We introduced the concept of a galvanic couple, which is a chemical process. Stainless steel is steel and comes in many chemistries or alloys. These are commonly grouped for fastener applications as austenitic, ferritic, and martensitic. It is called stainless in that rust stains typically will not form on a passive stainless steel surface. Chemically, the measure of how acidic, basic, or neutral a material is, is given on a numeric scale with 7 being neutral. Immersing some types of stainless steel fasteners in a solution of acid and water, mixed, and handled very carefully removes any surface iron from tools and other contact with fastener manufacturing machinery. This is called passivation and is a fastener chemical process. A working knowledge of chemistry is very useful for the practicing fastener specialist.

Reference

1. Jim Speck, P.E. *Fastener Fundamentals Seminar, National Industrial Fastener Show and Conference*, Columbus, OH, 1995.

6

Assembly Sites and Systems

6.1 Accessibility

One time or another, most mechanics have considered the thought, "I'd like to see the designer of this assembly work on it!" In the often frenetic pace of industrial activity and competition, it is all too easy to overlook the ability of an assembly's fasteners to be reached by someone performing work on an assembly.

An easy response is that the assembly will never need to be worked upon. Indeed, there are numerous examples of products that will never be unfastened. If that is the certainty in your application, then concern with reaching the fastenings does not matter. In reality, a majority of products do require some degree of accessibility. It could be that only a fraction of a percent of a production run will require work in their service lifetime. Which ones are they? Since this is unknown, it is prudent and value adding to provide access to possible service-requiring areas in any assembly.

With the concept of concurrent engineering taking hold in industrial product designs, the idea of "throwing a fastening design over the wall" between design engineering and manufacturing is no longer valid, let alone an efficient process of putting components together. And putting components together well and efficiently is what assembly is about.

Nor does it yield assemblies that have good access for farther along in the assembly's product lifetime. Let us compress time and have a technician working on an assembly that needs to be accessed for service conversing at the same time with a design engineer laying out the assembly on their computer. What would they have to say to each other? For certain, the access to the fastening system and specific fasteners would be high on the technician's list of topics. The service technician could walk the assembly designer through a service disassembly/reassembly sequence, using their experience and skills to perform each task step as efficiently as possible while illuminating for the assembly designer any part of product and assembly design. Steps that are either awkward and otherwise inefficient, or just should not be performed to carry out the assembly service, could be considered for redesign.

The technician might also ask why some component is positioned in the assembly in a specific location and together they could discuss other assembly sites that might be preferable from a service accessibility viewpoint. A good example would be the spark plugs in an engine. When initially installed in the engine out of the chassis, plug access is not an issue. Once assembled into the automobile, access might be severely restricted. Dialog with the designers and technicians can highlight this.

Another service technician observation could be the fastener service access decrease and performance loss with time and service environment exposure. This can be especially true of fastener drives and corrosion effects on mating fastening clamping surfaces such as threads and heads. The service technician might also be able to provide the assembly designer with competitive benchmarking information from experience with similar assemblies from other manufacturers. Both assembly access strengths and weaknesses can be useful information.

The designer can supply the service technician useful information regarding the proper approach angle for access to assembly fasteners. The designer can also highlight the size and directions of fastening forces while the assembly is in operation and while being fastened and unfastened.

The designer can also provide the service technician with useful information on assembly tools, both stock items and those of special design. In our previous spark plug example, the designer could suggest universal joints for the spark plug wrenches, which expedite the technician's task. The designer can also draft special assembly tools and fixtures, which can prove invaluable to the owners and those people responsible for servicing and maintaining an assembly. We can summarize some of the concurrent fastener access information, which should flow between the fastened assemble designer and the user.

Accessibility Factors	
The design engineer	The service technician
Fastener design	Fastener location
Type of fastener	Approach angle
Corrosion protection	Technician's tools
Competitive benchmarking	Assembly techniques and procedures

Concurrent dialog between those designing the assemblies and those who will work on them can yield more efficient fastening than by working independently. Figure 6.1 illustrates the balance between the assembly designer and the assembly's servicing technician.

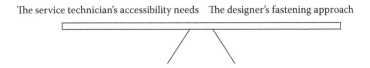

The service technician's accessibility needs   The designer's fastening approach

FIGURE 6.1
Design versus servicing balance.

6.2 Reusability Factors

As has been underscored elsewhere in this book, the prime performance measures of any installed fastener or fastening system are strength, appearance, and reusability. In this section, we will consider fastener reusability and the fastener factors that determine the assembly's long-term value as a mechanical product.

A listing of the fastening factors that have an effect on the ability to reuse a specific style or type of fastening would be difficult given the wide range of fasteners available. A partial list would include the following factors:

Damage to drive during installation

Corrosion in place of the fastening

Buildup of debris in fastener drive

Seizing of the threads from friction

Thread seizing from heat/corrosion

Damage from vibration

Metallurgical fatigue—metals

Ultraviolet degrading—plastics

Loss of locking ability

Thread wear

Bearing area loss

Service misplacement of fasteners

Let us look at these factors individually and consider some tips for countering any negative effects on fastener reusability.

Damage to Fastener Drive—A common occurrence with cross recess, hex head drives, and socket-type drives is for the tightening wrenches or tools to "cam out" or in other ways stress the fastener material through yield so the as-manufactured geometry is damaged. A good approach in these cases is to look at the geometry and material condition of the tools to make sure the geometric fitup is proper for the fastener type and the tool material is not soft on its working surfaces from metallurgical decarburization. Also inspect the statistical distribution of the fastener drive geometry to ensure that they are within IFI standards. On cross recess fasteners such as Phillips (so named for the original designer who benefited little from his work), Pozi-Drives, and the socketed types such as hex socket caps, commonly called Allen heads for their original manufacturer, a penetration gage can give the recess or socket depth. Figure 6.2 shows an IFI type 1 Phillips recess penetration gage.

Corrosion in Place of the Fastening—In assemblies which are used outdoors or in less than clean conditions, the elements found in the environment can cause a large reduction in fastener reuse with time. Elements such as sulfur, sodium, and chlorine combined with naturally occurring moisture can form hydrochloric and other acidic compounds which can corrode the working surfaces of a fastening, sometimes to catastrophic results. Some strategies for minimizing these effects are to choose materials which have some resistance to these elements, avoid combining dissimilar elements in an assembly to reduce galvanic couples and the resulting electrochemical potential differences which result, and use sacrificial coatings and the assembly parts and fasteners involved. Corrosion testing of the assembly and fastener materials and coatings prior to assembly production is often time and money well spent if it points up and helps avoid especially rapid corrosion factors. A final caution is to be sure that care taken in the design, selection,

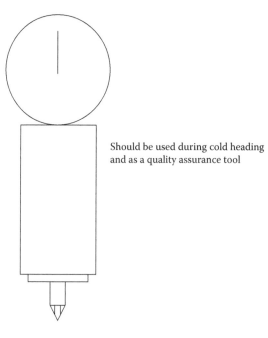

Should be used during cold heading
and as a quality assurance tool

FIGURE 6.2
Cross recess penetration gage.

and manufacture of assembly component, and fastener materials and coatings is not compromised by rough handling, nicks and scratches, and other pathways for corrosion to start in the protective coatings of parts. Often an initial and then periodic coating of light oil can provide help in this area.

Buildup of Debris in Fastener Drive—Any fastener which has a drive such as a Phillips, Pozi-Drive, Frearson, or Japanese Camera, cruciform recess, or a socketed type such as an internal six-lobed or hex recess can have grit, grease, and operating lubricants build up in the drive and reduce their torque transmitting ability. There are several steps which can be taken to minimize these effects. In fastener selection, avoid drives which are overly prone to camming out easily. Debris will only increase this tendency. A simple tightening torque test of candidate drives during fastener specification can pay dividends in future reusability under adverse service conditions. Next, locate fasteners where possible so that they are shielded from direct discharge and accumulation of debris materials. And finally, if they cannot be avoided, an application of protective grease or lubricant which can be flushed out of the drive when servicing is required can make the accumulation of fastener drive debris less of a servicing problem. In a coal pulverizer the author had been called in on, the socket head cap screw drives were ultimately filled with a potting compound to prevent a very hostile pulverized coal dust from rendering the internal hex socket drives of large socket cap screws unusable.

Seizing of the Threads from Friction—Thread galling and seizing can render machine screw threads useless by either requiring more loosening torque than the fastener drive is capable of transmitting or raising the thread coefficient of friction, K to a value of $K = 1$, where virtually all of the applied torque is consumed in friction. The threads can literally be "welded" in place due to friction forces melding together the high spots and surface roughness on the pressure planks of the threads. A convenient fix for thread seizing is to use mating threads of non-galling metallurgy or to separate the pressure flanks and bearing surfaces with appropriate lubricating and anti-seizing compounds. This can be particularly important for stainless steel threaded materials.

Thread Seizing from Heat/Corrosion—In threaded applications where elevated temperatures are service factors, mating thread pressure flanks as well as at the crests and roots can seize. This can especially become factors in applications with dissimilar mating fastener and site material and different thermal linear expansion coefficients. In these applications, adequate clearances should be used along with temperature appropriate thread lubricants and anti-seizing compounds. Figure 6.3 shows where thread seizing can occur.

Damage from Vibration—Loosened fasteners can cause reusability problems ranging from the hammering and fatigue of previously well-clamped parts to fasteners which shake completely out of the assembly site and are either lost or fall into working components with the further possibility of damage. An initial design step which can be helpful is to analyze the vibration modes in an assembly design and position the fastening sites to not

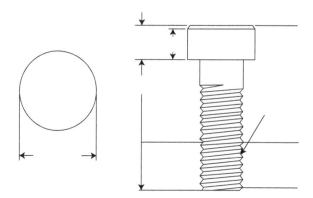

FIGURE 6.3
Thread seizing areas.

coincide with location of undesirable amplitudes. Also helpful with fastening for vibration is to employ damping, including fasteners where non-metallic fasteners and fastening features can provide elastic vibration attenuation. Self-threading fasteners, some with supplemental locking features, can provide relief from vibration-induced fastening problems. Features such as locking threads, both plastic inserts and coatings, commercial thread-locking adhesives and metal-locking features, and underhead serrations and locking washers can be evaluated. A good evaluation test is to measure the torque to install the locking fasteners, called the prevailing-on torque, the prevailing-off torque for the first use, and the prevailing-off torque on a fifth removal. The ratio between prevailing on- and off-torque, as well as the drop percentage between the first and fifth off-torque are good vibration resistance evaluation criteria, as are tests under vibration simulation such as a paint shaker type fixture prior to choosing assembly fastener sources.

Metallurgical Fatigue-Metals—It is axiomatic that fatigue can reduce the usable life of fasteners and fastener systems. Preload above the alternating service loads combined with making the assembled joints rigid with respect to the spring rates of the fasteners can minimize fatigue fastener problems. Additional steps are to know the endurance limits of the fastening systems and replace fastenings when their limits are being approached. Liquid penetrant and magnetic particle techniques inspection tests of assembles and fasteners can be used to detect micro-sized cracks before they can propagate and cause problems.

UV Degradation-Plastics—If prolonged exposure to sunlight is part of the service profile of a fastening system, evaluate its effects on any plastic fastenings used. Knowing their UV resistance and providing shielding were appropriate. Figure 6.4 shows the sun's work on a plastic fastener.

Loss of Locking Ability—Lock washers, locking tabs, thread-locking adhesive compounds, and a wide range of self-locking thread types can fail to deliver the reusability that is needed if they are installed or applied

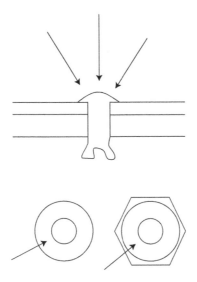

FIGURE 6.4
UV impingement on plastic fasteners.

improperly or in ways that minimize their performance. The operating principles of locking features and fastening products should be understood by both the engineers and designers of the assembly and the technicians and personnel who will use and possibly work on the assembly at some point in its life. Document and communication the locking features, their performance, proper use techniques and any cautions such as the need for thread cleanliness of thread-locking adhesives and burr-free intact surfaces for metal locking elements. An interesting approach to thread locking of machine screw threads is a tape which places the locking feature in the tapped hole rather than on the installed fastener. The internal thread pressure flanks have a secondary locking ramp rather than the standard 60° included angle. On installation, the external thread pressure flanks are elastically deflected slightly to provide higher friction and are more resistant to loss of locking ability.

Thread Wear—Wear and tear of machine screw threads by the tightening and loosening of high-frequency threaded assemblies can be minimized by judicious selection of mating thread materials and finishes. A relatively abrasive zinc plating on steel fasteners can cause undue friction, which can cause the sliding thread contact surfaces to gall and wear down at the high spots, thereby diminishing thread fit. Harder finishes such as electroless nickel can add both the hardness plus lubricity to minimize wear from frequent use. Additionally, wear from cross-threading and mis-starts can be minimized by thread starting chamfers, pilot, and full dog pint and other locating features.

Bearing Area Loss—Loss of bearing area from distortion and compression can be avoided by calculating thread bearing area stresses. Balancing the assembling bearing strength and fastener area can avoid bearing area

FIGURE 6.5
Typical fastener bearing areas.

loss. This can be accomplished by understanding the bearing strength of the component materials and the fasteners along with the fastener's footprint on the components. I often use the analogy of a pair of snowshoes to demonstrate the bearing area phenomenon. If the bearing strength of snow (or a component) is below a given strength, a person would sink when walking (or clamping) on it in normal footwear. By increasing the bearing area with snow shoes, bearing stress is lowered and walking (or effective clamping) is carried out. Figure 6.5 shows two typical bearing areas.

An example of bearing area stresses would be

Head diameter 0.250 in.

Thread #6-32

Fillet radius 10% of shank diameter or 0.10 (0.138) = 0.014

Clamp load, 1200 lb

Inner diameter is comprised of shank and twice the fillet radius or 0.138 + 2(0.014) = 0.166 in.

Finding the bearing area:

Outside diameter area − inner diameter area

$0.7854 (0.250)^2 − 0.7854 (0.166)^2 = 0.0274$ in.^{2}

This is the fasteners effective bearing area provided the bearing surface is perpendicular to the axis of the shank. Two percent is a typical tolerance for bearing surface to shank perpendicularity. The more this is consumed in manufacturing the fasteners, the higher actual localized bearing stresses may actually become.

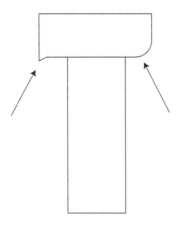

FIGURE 6.6
Localized bearing stresses.

Bearing stress is found by dividing clamp load by bearing area:

$$1200/0.0274 \text{ in.}^2 = 43{,}723.8 \quad \text{or about } 44{,}000 \text{ psi}$$

Figure 6.6 shows the results of high localized bearing stresses. If the assembly compressive yield strength were lower than 44 KSI, say if the top clamped assembly member is wood or low-modulus thermoplastic, than permanent compression could occur with a loss of clamp load and possible assembly damage. Here an increase in bearing area by the use of a larger bearing area fastener or a washer might be in order. Also bearing perpendicularity would be important to avoid localized raised bearing stresses.

A good diagnostic of reusability performance is to periodically audit the feature or features of the fastening system which have the most vital influence. For example, if molded-in snap damage or less than targeted fit is a prime factor, perhaps once a month pull some assemblies off the line at random. Evaluate the fit. Document and chart the fit. And analyze and communicate to all team members the root causes for any less than desired performance. Do this with the goal of a steadily increasing trend in fastening performance.

If wrenching surface distortion is a factor affecting reusability, a failure modes and effects analysis can highlight both the ways this fastening performance loss occurs and its subsequent customer impact.

Often gages and other special-purpose, but simple measuring equipment can be constructed or obtained to monitor and control important fastening system reusability features. For example, the cross recess penetration depth is a prime factor in the future serviceability of Phillips and other cross recess type fastener drives. By obtaining a simple recess penetration gage such as the one shown in Figure 6.7, recess depth can be kept track of with the intention of only sourcing fasteners with recesses to a specific controlled depth.

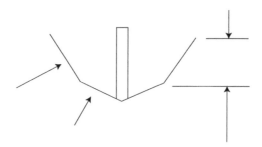

FIGURE 6.7
Side view of fastener cross recess.

Without such a measurement statistic, this depth can go uncontrolled with no indication save the fall in reuse performance. Figure 6.7 shows a side view of a cross recess indicating set and point angle and the penetration datum which is needed along with proper fitting and condition drivers for good torque performance.

Figure 6.8 shows a typical assembly tool for semi-automatic hand use.

Misplaced Fasteners—If designed-in fasteners are removed from service and set down in an area which is unplanned, the possibility of loss and reassembly with fasteners missing or inferior substitutes can lead to a diminishing of assembly performance. By providing a temporary storage nook for fasteners during service, the reusability of the assembly is enhanced. Figure 6.9 shows an example.

Figures 6.10 through 6.13 show some typical automatic fastener tools and installation equipment.

FIGURE 6.8
Typical assembly tooling.

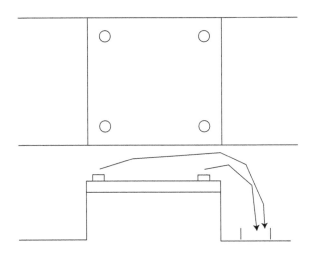

FIGURE 6.9
Fastener temporary storage area concept.

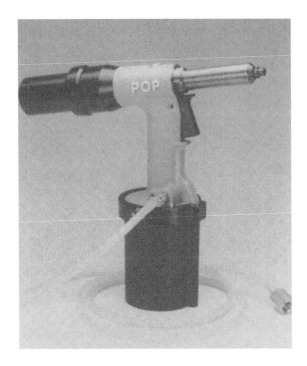

FIGURE 6.10
Automatic fastener equipment.

FIGURE 6.11
Fastening equipment.

FIGURE 6.12
Automatic fastening machine.

FIGURE 6.13
Fasting equipment for volume assembly production.

6.3 Assembly Training

It is perhaps unfortunate that the old saying "We never have time to do a job right but always have time to do it over" holds true in the assembly business. It is in the training of assembly operators that the dramatic gains in productivity and efficiency can yield true costs benefits in the form of quicker assembly times, lower production costs, and an enhanced competitive position in the marketplace.

To watch the demonstrated value of assembly training in motion, observe the studied, rapid motion of a top race car pit crew in action during any major automobile race. Each movement by every team member is performed to a well-practiced plan, with no lost or random work being expended. Every assembly job to be performed on the car, both the scheduled removal and replacement of wheels and tires, suspension adjustments and refueling, as well as unplanned operations such as the repair of damaged body parts to control airflow and replacement of engine components are performed with maximum speed and minimum confusion and lost motion. To what can this assembly efficiency be attributed? It is training. Each mechanic knows in advance what task to perform and the best way to perform it. The most efficient task sequence of steps and motions as well as the availability and positioning of all needed tools and components are a common body of knowledge for the crew members. Even unscheduled operations are handled efficiently. A good analogy for general fastening work would be the special order or engineering prototype run. So it should be with any production assembly operation Assembly tasks, skills requirements, and site layout should be studied and practiced numerous times prior to the "race" of an industrial assembly operation so that in production their execution is routine.

If efficient assembly operations are to be a reality in your work, training needs to be a key component in the planning and implementation of the people doing the assembly work. Assembly training can take many forms and can be tailored to best suit the requirements of your specific application. In many companies and countries, operators receive on-the-job apprenticeships and training in the basics of assembly mechanics and procedures. Using the race car business model again, many operators start new employees in the race shop with gradually increasing task responsibilities before finally moving them into a place on the pit crew at which time their training is well established.

6.4 Degree of Automation

When making decisions about the scope and degree of fastening equipment and automation to employ in your application, a useful model to use is a simple input/output diagram. Using this technique allows you to match the investment in automation with the assembled units being demanded by the assembly operations customers. This also has the advantage of providing for periodic reassessment. In many real-world manufacturing situations, the number of assemblies required over the life of a product line varies, often starting with relatively small pilot lots and a low assembly automation requirement. The demand can explode exponentially during successful growth periods. Some facility may finally have to be made for assembly production during product maturity phases when it is still profitable and prudent to produce assemblies, however at a much lower rate of assemblies per unit time. Lower rates may also be called for as an assembly builds flexibility for limited or intermittent production runs of "special" lower volume versions of the standard high-volume assembly. Making an input/output model helps lay down the framework for these decisions.

Let us look at a simple example and assume we are going to assemble an electrical connector for an automotive, under-dashboard component. Let us further state that each connector assembly is very efficiently designed with one blind rivet accomplishing the joining of the three separate component parts.

Our production forecast calls for 300,000 connectors annually. The output side of our simple model is shown Figure 6.14.

Using the output requirement for our assembly operation, we can put a time constraint in assembly time per unit.

Time available

Let us state 230 days/year × 8 h/day × 60 min/h

Time available = 110,400 assembly min/year

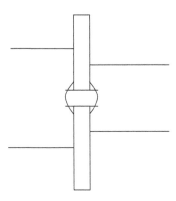

FIGURE 6.14
Fastening output.

Time unit

300,000 connectors/year/110,400 min/year = 2.72 min of assembly time per connector

So our three-component inputs need to be put together and riveted in 2.72 min, or one about every 163 s. This one-every-seven-minute target gives us the mean time assembly target (MTAT). Our assembly procedures, tooling, hardware, and process operators must be designed to consistently assemble at this rate or better with some margin of safety. Using an MTAT of 163 s, we can diagram the times required for each step of our assembly operation as a total, then break it down into fractional times for each step.

Let us diagram our assembly steps:

First, we have to bring together, orient, and position our three components shown in Figure 6.15.

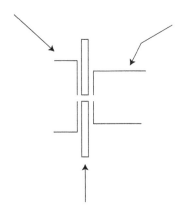

FIGURE 6.15
Three-assembly components.

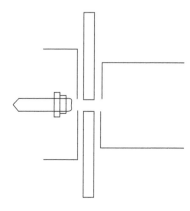

FIGURE 6.16
Picking and placing the rivet.

Then, we have to pick and place the blind rivet, as shown in Figure 6.16.

We then have to pick and place the riveting tools and rivet. Finally, the completed assembly is exited and the assembly cycle is repeated, as shown in Figure 6.17.

If we analyze the steps, we can categorize them as transit steps and fastening steps. In our example, we have three transit steps—components transit, rivet/tool transit, and exit—along with one fastening step—rivet. This gives us an idea of how much fastening to preparation exists in our process.

If we diagram the assembly cycle along with the mean time assembly target, we have a good diagnostic to graphically represent whether we are assembling within our assembly target. As we see in Figure 6.18, we are. If we only timed to cycle without breaking it down and diagramming it, we would know only the rate and not the shape of the assembly time function. Figure 6.18 shows the time function graph.

We can also see from the diagram that transit time takes a large percentage of our total unit assembly cycle. This gives us useful information for cycle

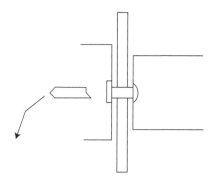

FIGURE 6.17
A full assembly cycle about to be completed.

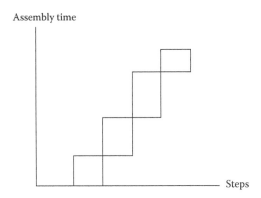

FIGURE 6.18
Assembly time graph function.

improvement within our given assembly process. We can use this tool to serve as a metric for assembly process output improvement. Writing individual step times, we have

1. Component transit(s)	20
2. Rivet/tool transit(s)	25
3. Rivet(s)	8
4. Unit exit(s)	10
Total unit time(s)	63

Some useful ratios can be evaluated from this data. Our percentage of unit assembly time/MTAT is 38.6%. Our assembly process meets our assembly target with a margin of 1.4. Evaluating the ratio of transit time to fastening time, we have the following:

Assembly 63 s/Mean Assembly Time 163 s

If we first look at the fastening time, we can assure ourselves that this is a reasonable rate and look for any riveting motions that can improve on this time. Next, we can look at the large percentage of the unit assembly time consumed in transit. Here, we have fertile ground for improvement with a positive impact on our assembly time and target margin. Perhaps a change in the relative motions orienting and getting the three components in position for riveting can be accomplished. It could be that placement of the three components requires the assembler to reach for components and consume time. The goal should be to reduce these transit times. The transit work performed is of a less value-added nature than the actual fastening. Reducing the transit times and improving the ratio will yield a more effective time function diagram. More importantly, it will result in more efficient and

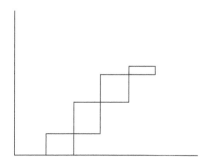

FIGURE 6.19
Improved transit motion.

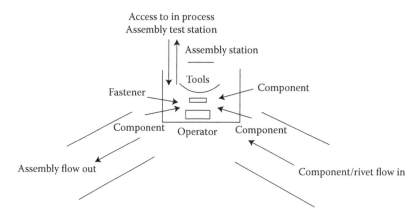

FIGURE 6.20
Assemble station layout sketch.

profitable assembly operations. Sometimes, the development of assembly bins, tooling, and fixtures can greatly reduce these transit times. Let us look at a unit assembly diagram where we have improved our transit motions, as in Figure 6.19.

Figure 6.20 shows an assembly line layout sketch for a typical assembly station.

6.5 Automatic Assembly Machines

When assembly build quantities increase rapidly, as can happen when the product an assembly is in meets with commercial success and demand success, a logical step is often to have an automatic assembly built and installed to ramp up assembly production. Whether this ramp up occurs in a timely manner and to everyone's satisfaction is dependent on the automatic assembly machine planning, decisions, and implementation.

The decision regarding an automatic assembly machine or bank of machines is similar to excavating and pouring the footings and foundation of a house. If these decisions are done poorly, no heroic efforts that follow in assembly operations can restore all of the possible assembly efficiency gains not realized. Planned and executed well, an automatic assembly machine or machines can be a logical and profitable assembly management response to an increase in assembly demand.

Rather than accepting this on faith, let us expand our previous automotive assembly example. Let us say the build rate has grown considerably due to expanding demand. We are now building over 10 million connectors a month, with some months requiring a 1 million assembly per month assembly rate. Using the same model, we now have to build at the following rate:

Use 20-day month and 8-work-hours as conservative figures

1,000,000 units per month/20 days per month = 50,000 units/day

50,000 units per day/8 h/day = 6250 units/h

6250 units/h × 1 h/60 min × 1 min/60 s = 1 unit/1.73 s

We now need to assemble a unit every 1.73 s

Our MTAT is now down to 1.73 s per unit if we want to use an automatic assembly machine. If we propose instead that we merely duplicate the number of assembly stations of our previous assembly layout, we would have four stations if we maintain the same margin. While this would give us the flexibility to put on and take out of service stations to meet demand, we would be adding an operator at each station. This would give us a high labor cost. Laying people off to react to a reduction in future demand would also exact a cost to both the assembling company and to the operators involved. An automatic assembly machine will let us meet the new MTAT while still allowing us to maintain a more controlled number of assembly operators. The operator, or operators, will now control the automatic assembly machines operation rather than performing the assembly and fastening. Once the decision is made to obtain an automatic assembly machine, it is important for the decision maker or makers to first know firmly their mean time assembly target. Next, a rigorous collection of information should be made on the options available along with the strengths and weaknesses of each. A good starting point is by doing competitive benchmarking of assembly companies in similar product categories.

In our example, it may be unreasonable for competitive reasons to expect a company making a competitive connect to share automatic assembly information with us. But we can tap into all high-volume assemblers fastening with blind rivets. While their components will be different, several of the automatic assembly machine stations, including the critical blind riveting station should be similar enough that information valuable to our automatic assembly machine decision can be gleaned. A good source for benchmarking contacts would be our blind rivet supplier, who has a vested interest in our continued success with the assembly. When making the decision to obtain an automatic assembly machine, knowing the experienced assemblers and learning from their experiences can help avoid mistakes others have made and help position your move to automatic assembly for your positive competitive advantage. As an example, they may offer that if they had to do it over again, they might add an additional station to provide an operation not originally anticipated. Once the machine was built and on site, adding the additional station was much more expensive. These well-learned lessons can prove valuable in your assembly machine decisions. Once the information is starting to come together, a decision regarding the basic layout of the automatic assembly machine must be made. It is helpful to break this down into specific layouts for the purpose of organizing the decision process. Four basic configurations of assembly stations are as follows:

Rotary table	The stations are positioned around a circular table
In line	The assembly flows through the stations linearly
Horseshoe	The assembly process starts and stops at adjacent stations
Oval carousel	The stations make a circuit around an elongated assembly process path

Each has its advantages and drawbacks. None is universally right for every operation. The factors that determine the best automatic assembly machine for each application will carry widely varying weights. While one application may require wide spacing for station access, another application may require a smaller machine footprint due to floor space constraints. Their throughput speeds may meet, or be outside our MTAT time. In our example, we would be looking for a throughput time of under 1.73 s.

Some factors that should be carefully evaluated in selecting an automatic assembly machine are presented in the following text.

6.5.1 Station Tooling

Good quality assembly tooling such as grips, fixtures, driver bits, and clamps should be part of the initial machine set up and established as ongoing assembly machine supplies. Be aware of the type, hardness, and finishes of both the initial and replacement tooling so that the planned-for throughput speed is consistently delivered by the assembly machine. For example, cross-recess driver bits can be obtained in a good, better, and best set of material conditions. Spending a little more for the better or best assembly tooling materials can save money in assembly machine downtime.

6.5.2 Ergonomics

The automatic assembly machine should be well suited to the assembly site and provide ready access to all needed areas of the machine. Stations to be frequently accessed by operators, in particular, should be designed with human factors in mind so that operation of the machine will be low in operator fatigue. Having operators visit or talk to current users of a specific automatic assembly machine can yield benefits in future operator acclimation and efficient use of the intended assembly machine and its stations.

6.5.3 Designed-In Flexibility

A key is the future ability of an intended automatic assembly machine design to have the flexibility to change over to new designs or related assembly production should the need arise. This can be accomplished by using assembly stations with industry common station drivers and ancillary equipment such as riveting heads, parts feeders, and conveyance systems. This increases the final recovery value of the automatic assembly machine and the marketability of the unit should business conditions warrant.

6.5.4 Seamless Handoff

While the tendency is often for an automatic assembly team to want to move on the project once a design is selected, it is time well spent to cement the relationship by having many involved members from both the using and supplying companies communicating so that the automatic assembly machine can be turned over from design and build to operations with a smooth flow of know-how and past experience. This will increase the early confidence of the machine operators.

6.5.5 Setting Quantitative Targets

Meshing with the assembly MTAT, a well-distributed log or chart should be maintained so that assembly speeds are kept on course.

A final note, do not cut corners on planned maintenance as time passes. Good maintenance maintains assembly performance.

6.6 Smart Machines and Robots

A trend that shows no sign of slowing or reversing is that toward increased automation in high-volume assembly operations. This has profound implications for the fasteners and fastener systems that are used at highly automated assembly sites. The automation trend, coupled with a competitive consumer market requirement for assembly flexibility, leads logically to the use of industrial robots and smart assembly machines for the performance of repetitive assembly operations.

If the build volume and added value of the assembly warrant it, using programmable robots and programmable smart assembly machines can accomplish the twin goals of reduced assembly unit costs and increased flexibility, and reduced assembly change over time. Two examples are offered.

First, a leading manufacturer of a line of high-volume consumer electronics products required the capabilities of an efficient assembly facility, which could assemble and fasten with precise positioning an electro-mechanical precision mechanism in a complex injection-molded chassis. Being located in an area where labor costs were high, but not wanting to have the assembly carried out to a remote location and lose direct interface, this manufacturer designed and had built the entire assembly system with a series of programmable robots. Each robot was fitted with end effectors, or hands, which made them extremely capable of repetitive, high repeatable precision installation and placement of each of the miniature components in the assembly. The assembly room in operation was a whirl of 100 robots carrying out their tasks as conveyors handed off each successive step to a robot, which rapidly indexed to its task, while two technicians monitored assembly production.

What was observable was the high initial cost of the robots. Also notable was the efficiency of the robotic assembly line as it seamlessly sped through, what would be for a person, very intricate hand movements. Other observations were that while many individual operators were not required as with a traditional semi-automated assembly line or automatic assembly machine, a new requirement with the robotic system was the need for well-trained technicians to program and monitor the robots. And the robots were less discriminating in noncompliant components or in component shipments from suppliers that contained foreign material or duds such as bent stampings or unthreaded fasteners.

The second installation was for a much larger and heavy duty assembly, the connecting rod assembly for a 13-L diesel tractor engine. The con rod assembly contained the rod and rod end, crank bearing and wrist pin bushing, and two large-diameter, highly engineered connecting rod screws threaded into holes tapped with a thread-forming, as opposed to a thread-cutting, tap. An interesting fact about connecting rod bolts or screws is that the big end can be subjected to wide changes in load and load direction with engine operating conditions. The highest service loads on these threaded fastener joints occur when a driver takes his foot off of the accelerator, slowing the flow from the diesel fuel pump, while the pistons dynamically brake the tractor, without the benefit of a full-combustion charge in the cylinders. As we learned in previous chapters, preloading is the proper method to minimize service load fluctuations in these relatively rigid joints. A microprocessor-based, torque-transducer-driven wrenching system was the heart of this intelligent system.

The microprocessor was capable of analyzing both the torque on the transducer and the degrees of rotation made by each connecting rod screw as it was being installed. By comparing it mathematically with the torque-rotation function, or signature of a conforming, properly preloaded connecting rod joint, the smart assembly system was capable of differentiating between normal process variation and special causes such as a bolt material too low in yield strength or cross threading on starting.

This analysis of assembly operating conditions by the assembly machine allows the machine to function as a thinking, decision-making unit. Assembled units that are outside of the preload specification can be identified for rework and repair.

Some unique points about a smart system and the use of robots and microprocessors is that data about each assembly can be stored for future reference. The connecting rod screw suppliers can be evaluated on the performance of the fasteners they ship using data right from the assembly machine, so long as it is programmed correctly.

Assemblies not meeting engineering specifications can be reviewed downstream so that the factors causing noncompliance can be controlled. Best of all, in a high value-added assembly such as connecting-rod assemblies, the preload record is available as positive objective evidence of the production of a quality assembly should the need arise in the future concerning the performance of a unit delivered to and put in service by a customer.

Factors to consider for robotic/smart machine assembly:

The technical training requirements for personnel working with the assembly robots and microprocessor-based machines.

The need to either have as close to 100% specification-conforming parts and fasteners as possible or screening stations for culling out foreign material and duds, which might confound the robotic system.

The need to provide for a system of rapid reprogramming of the system in cases where an assembly design change is to be made.

The possible need for an offline training and programming station where technicians and assembly designers can increase their programming skills proficiency and carry out prototyping assembly work.

The need for end effector and fixturing preventive maintenance to protect against wear-induced loss of assembly alignment and precision.

The sea change that changing from more traditional assembly line processes will represent and the impact on the assembling company's culture and its design rules and protocols for future assembly projects.

The changed requirement for assembly site environment and layout area access and environment may require changes in the way the assembly area is maintained.

Assembly sites and systems.

6.7 A Renewed Look at Modern Assembly

Manufacturing has seen a renewed vitality and growth in both North America and a continual expansion worldwide. This is certainly a welcome trend. Training new employees, some of whom may be new to manufacturing and assembly work, continues to be a challenge. It is one that engaged company managements can address and solve with the help of education, community, and government leaders. Two points that will help start this update to the first edition is the very concept of the team and team member. The other is a story I heard on the car radio while on the way to a business meeting this week. The news story first concerned missing rivets in automobile air bags. This touches the subject we are addressing here about fasteners, assembly sites, and systems. Shortly after a news report, a recall was issued by a leading pickup truck manufacturer involving hundreds of thousands of vehicles due to a fastener in the rear axle assembly, which could loosen. Fortunately, no injuries were reported although accidents were involved. The recall involved fitting a retaining device to secure the fastener we define success of an assembly operation as the output of efficient assemblies that we provide safety and integrity in operation and minimal variance in projected cost. To achieve this goal, assembly sites and systems must be designed to achieve this goal and be operated by people, equipment, and materials with the requisite skills, motivation, and capability. If successful, designed experiments and test metrics will indicate to the trained observer that they are on target.

A good assembly operation model is the race car pit crew. Team role members would include a technical manager, perhaps an engineer by training, of course a team owner, team members with assigned functions, and also importantly, someone handling the financial functions, perhaps an accountant. As a backdrop, rules and legal constraints would be involved and would be in the sphere of influence of an attorney. Let us compare some of these team functions.

6.7.1 The Engineer and the Accountant

An accountant carefully studies the sources and uses of funds in the organization. The accountant also watches the flow of cash on a frequent basis. By knowing where the money is coming from and where it is going, and the rate of cash flow and cash flow changes, the accountant helps ensure an adequate supply of cash at hand. Cash availability is liquidity. And cash is the life blood of a commercial operation, including assembly teams. Run out of cash, and the team is out of operation. This is called going broke, as opposed to going for broke.

The engineer is a creative person. Good engineering design, including design for assembly, is both a science and an art. The engineer defines a problem and uses science to artfully, and if done well, efficiently solve the problem. The engineer wants to solve problems with processes and products that are operated and assembled by team members for the greater benefit of the team and the organization to which it belongs. The pit crew that succeeds, puts its car and the driver in the best position to win the race and be that season's champion. For an assembly team, it is to assemble items that can be sold at a profit. Profit is necessary for the success of the business.

Both the accountant and the engineer have the goal of profit and success for the business. Their timing with respect to cash is almost always out of phase. The accountant wants cash up front and a positive cash flow. The engineer wants to create the products and processes, test and refine them based on careful analysis of test results to design targets, and then launch a new or improved product or process.

The conflict in the short-term, day-to-day goals of the engineer and the accountant are based on the timing differences with respect to cash flow. The engineer's work, say a new assembly system, will consume cash for some period of time before the work results in output, which can earn the organization profits. The cash flow is negative at the front end and then turns positive if done well.

The accountant has the fiduciary concern with assuring the firm does have an adequate reserve of cash and has liquidity. The accountant's concern is that cash should always be positive. The solution to avoiding conflicts on assembly projects is clearly stated goals for both the accountant and the engineer. Time lines and milestones should be clearly documented and quantified. For the accountant, clear cost and time limits should be given to the engineering personnel with responsibility for managing the assembly project. For the engineering group, responsible adherence to the cash flow parameters is an important part of the project. The engineer's natural interest in concentrating on the technical aspects of the assembly work cannot be allowed to override the economic parameters. With clear and open communications between engineering and accounting, assembly projects can be both technical successes and provide economic benefits to the companies who employ them.

6.7.2 The Engineer and the Attorney

All fastening and assembly commercial operations work within a legal environment. The attorney provides the legal insight into the application of civil and legal laws to the engineer's interaction with the market. The engineer is working to provide the maximum fastening performance at a competitive price while providing the company with profitable operating profits. The laws which apply can be from the purchase orders and contracts for specific fastener and assembly products and services. The laws can also be from more general laws concerning commercial activity.

As an example, consider the purchasing contract for an annual quantity of threaded automotive suspension components being manufactured by one company, purchased by a second company and installed in a suspension assembly which is sold to an automobile manufacturer. The automobile manufacturer in turn sells the new car to a car dealer who then sells the car, with threaded component installed to a customer. If, in a year's production, a quantity of threaded components is not up to specification, what are the legal factors involved? The attorney and the engineer must have an understanding of what the proper technical procedures and documentation, which satisfies the applicable laws involve.

Let us consider other assembly team member roles and responsibilities. A good starting position is the team member. This individual executes a predetermined set of operations in the assembly process. They are working in, as opposed to on, the assembly process. Two important factors that can help determine the degree of success is the clarity of the operation's procedures, training and instructions, and the degree of individuality, which empowers this team's member. All are critical factors. Let us discuss them. Once the decision has been made to introduce a new team member to the assembly team, a smooth ramp up will often depend on the attainment by this team member of the skills required to carry out the assigned assembly tasks in a quality manner. A tried and true method for this type of training is the watch-try-measure-do method. The new team member watches the tasks being performed by the trainer. Next the team member performs these tasks with the trainer coaching the team member. This is the time to correct any observed departures from good procedure for these tasks. Measurement metrics such as seconds per task provide the quantitative tools to assist the new team member and coach. Measurement against the assembly metrics marks progress toward full integration into the team. Finally, the team member "does" operating as a full team member with periodic audit, review, and coaching by the trainer. Empowerment occurs when the new team member, fully functioning in ongoing assembly operations, takes ownership of the tasks. Consider any new physical task one could relate to: an infant learning to walk, a teenager learning to drive an automobile, or a person at any age learning to play a musical instrument. All of these require the learner to master specific physical skills prior to joining the team of those who went before and who now practice those skills. There comes at a point when the new individual has learned these tasks to a level where the tasks are internalized. The areas of the brain dealing with subconscious thought carry out the required thought processes without the individual needing any longer to think directly about performing step two after step one. In some famous assembly lines, empowerment provided the means for individual team members to stop the line if a nonconforming assembly was detected. As production efficiency is a primary goal of assembly operations, stopping the line is assured of focusing organizational attention of the assembly quality issue initiating a team member taking such a step. Less drastic team member empowerment techniques that have been used include the authority to identify specific production units for rerouting to repair stations to remediate the identified items. Considering measurement metrics for assembly teams, units assembled, in total, and per unit of time are fundamental to the assembly team function. Others which can prove useful are energy usage per assembly, rate of changes in assembly times, and inventory costs. The on-time delivery to promised and scheduled ship dates will be familiar to most readers with factory or manufacturing work experience. That brings us to another key team member, namely management. In most commercial assembly operations, an individual or group of individuals will have management responsibility, which often includes profit and loss responsibility. While the team member is properly concentrating on the assigned assembly task and the coach focused on the performance of the trainee, the manager's field of responsibility encompasses the entire process. It is the manager who can communicate the data of the measurements to those team members whose work exceeds or lags the numbers needed. In well-performing teams, this communication is open, non-punitive, free flowing, and immediate. A collaborative effort of this level requires time and effort but is usually well worth the effort. The assembled products of these organizations can build strong brand preference and loyalty in its customers. Attracting new team members becomes an easier task since it is human nature to want to belong to winning organizations. Let us take these team communication points individually. Open means that each member, coach, and manager can express their viewpoint without fear of reprisal, either overt or more passively but no less toxic to good team function. Nonpunitive means that "shoot the message" is neither practiced nor tolerated. It is remarkable how many operational malfunctions have occurred and were observed developing but not communicated in time to prevent a catastrophic event due to reluctance to speak up. Free flowing means that speaking, and more importantly, listening is always happening and can go from daily comments to high-impact calls to action seamlessly. And immediate speaks for itself. Areas for assembly process improvement are addresses in real time. Closing this section, it is important to bring in two voices. They are the voice of the customer, and the sound of the market. Sales and marketing can play an important, if indirect, role on the assembly team by being the customer's representative to the team assembling the customer's order. This important input can lead to key improvements to the end product. It can also bring attention to effort by the team that is expended on work and product features, which are not of commensurate value to the customer. This can help eliminate assembly effort and resources, which are being inefficiently being expended. Throughout my career, but especially early on, I discovered the value, if managed wisely, of interaction of the customer and the shop floor folks. It helped put a face on both the customer and shop personnel. As was once expressed by someone whose business skill I greatly respect, "People buy from people". This needs to be managed well and a skilled sales person can gain from mastering these interactions. The assembly team can also be empowered to greater levels.

Finally, no market is static. Customer's needs and expectations, offerings from motivated competitors and shifts in social patterns all work against long-term growth of a stable assembly. Mobile phones evolved as did main frame, desktop, and laptop computers. Early in the industrial age, the Model T was pressed by more modern automobiles. This pattern is ongoing. Input into the assembly operation is important to prevent stagnation. Flexibility in both team skills and process equipment is critically important if any specific assembly operation is to maintain its relevance and profitable output and operation in an ever-changing market place of ideas, products, and services.

6.8 Fastener Numerical Methods

It is fitting that in a chapter where we have discussed assembly teams, we introduce and review numerical methods with an emphasis on fastener and assembly application. In training classes I have often used the expression, "Words are good for description but numbers are better for measurement." As an example, if in response to the question "What time is it?" one can only use words, we are left with, early morning, slightly after dawn, or some similar answer. Using numbers we can easily say, It is 6:48 AM and be communicating clearly and accurately.

6.8.1 General Rules for Numerical Proficiency

Equal must be equal. There is a purity about math that makes it a useful tool for the fastener specialist. In a way, presenting numerical methods for the fastener specialist reader presents a unique challenge. A good analogy would be writing about playing golf or learning to play a musical instrument. Beyond the basic concepts, only good quality practice, and lots of it will get the reader to whatever level of numerical competence they set as a goal. Another challenge is mind set. Most of us have heard the comment "I'm not good at math." I have noticed in some of those declarations an underlining theme of dismissiveness, implying that it is not important or I can always get someone to do that number crunching. If you want to be competent technically in any engineering field, including fastener applications, mathematics is the solid foundation and structure upon which support all of the other topics. The old saying, "Show your work" can be subcontracted to other people.

You will also always be reliant on them. If their work is incorrect, you may not know. If you can "do math," you are in a stronger position. Knowledge in this area is real power. So let us go through numerical methods.

I will provide some thoughts and an overview. You are encouraged to pick up some problems from this text and make up your own. Practice working with them often, search out good collaborative help when needed, and always keep "working the problem at hand."

6.8.2 Numbering Systems

We use the algebraic numbering system of positive and negative numbers and zero. There are other numbering systems to know. Binary numbering systems use one and zero. It can be thought of as on or off. Binary logic is used in computer code and electronics circuits. Both the algebraic and binary systems have some rules that must be followed for these systems to provide reliable, as in true, results. In the algebraic system, some operations are undefined, such as division by zero and the square root of a negative number. To handle this last rule, a numbering system of imaginary numbers was developed denoted by the letter I to facilitate making these calculations. Similarly, the system of logarithms is widely used by scientists and engineers to calculate very large and very small numbers, the symbol Log denoted the base of 10 and with other bases such as for natural logarithms also in use. For fastener work, get to know the units of common terms such as areas, loads, and torque. An important numbering system rule is that "equal must be equal." When solving, be able to take anyone through your mathematical work and prove that the equality holds and the units remain consistent.

6.8.3 Geometry

Fasteners are rich with geometric shapes whose areas can be calculated. Some are easier such as the area of a circle and a square. Circles are found on fastener shanks and headed blank diameters. Some fastener drives use a square recess or socket. Others such as the hex, oval, cone, and others have additional factors. You just need to start calculating them. Know that geometric calculations are approximations. The closer the dimensional factors such as diameter and height are to the actual part, the more accurate the calculation will be. Areas will be in square units, metric or inch system units raised to the second power, volumes are cube units meaning numbers raised to the third power. Volumes, in fastener application engineering work, are very useful. Multiplying part volume times the fastener material's density gives us the part's weight.

6.8.4 Algebra

The use of symbols such as S for stress, P or L for load, and A for area are frequently found in the algebraic equations of fastener applications work. Algebra provides a language that is more precise for working on a fastener application. Equal must be equal. If I have pounds or newtons of load on the left side of the equal sign in $P = S/A$, I must find pounds per square inch divided by square inches or newtons per square millimeter divided by square millimeters on the right side of that equation. The equation must be equal and the units consistent. That is a good proof of the work. I found that working through chemical equations useful in practicing balancing equations and keeping units consistent. Another thing that helped is that early in my fastener career, another individual and I agreed to discuss our fastener work where possible in only mathematical terms for 6 months. It was difficult at first but became very useful. You had to have really thought through and know you statement before you expressed it. That is a good way to improve your mathematical skills. It is also how graduate students demonstrate their learning by defending their thesis to a panel of professors.

6.8.5 Trigonometry

Trigonometry is often used to solve fastener problems such as finding the diameter of a countersunk head or finding the across-the-corners dimension of a hex head. Trigonometry takes on real meaning and can be divided into right-angle problems, meaning triangles with a 90° corner and other angles. Much fastener trigonometric work involves right triangles. Get to know the ratios for the sine, cosine, and tangent functions using the side adjacent (a), the side opposite (o) and the hypotenuse (h) so that

> Sin Angle a = Side o divided by Side h and its solution becomes
> second nature to you.

For non-right-angle triangles, the Law of Sines and the Law of Cosines provide tools for their solution.

With trig problems, the importance is placed on determining what sides and angles you know, perhaps by part drawing or by measurement, and what trig function is required to solve for the unknown item of interest. For right-angle calculations, I have found a very simple technique to remember: O/H, A/H, and O/A for the sine, cosine, and tangent functions. I would write the letters OHAHOA vertically on the sheet of paper I was calculating on and the sin, cos, and tan to the right between each pair of letters. I would look at what two sides and angle I know, saw head diameter, head angle, and head height, and select the trig function that allowed me to solve for the unknown.

6.8.6 Statistics

This area is especially useful in quality control work. Important concepts are the population as a whole group, or lot size, and the sample; was the sample randomly selected or was there bias? and calculation of the mean and standard deviation of a data set. Being able to assess the scatter in a group of similar numbers is useful. Statistics is a very useful fastener tool as we often do not have the time or luxury to collect more than one sample from a lot of fasteners and make a decision concerning their conformance to specification. Also in business, using statistics to assess probabilities of success, say in which of two regions in which to make sales calls, can prove extremely valuable. The mean is found by adding up the numbers and dividing by how many were used. This is the same as the average. Sample mean and population mean have a slightly different divisor. In fastener quality assurance work, it is the sample mean that we will be using most frequently. The median is the central number in a distribution. The distribution can be a normal distribution—this is the familiar bell-shaped curve—or the distribution can be skewed. Punch wear and the corresponding recess-penetration depth are examples of a skewed distribution. The mode is the most frequently occurring value. Mean, median, and mode are key statistical terms. Variance is how far any individual member of a sample, or population, varies from the mean. Standard deviation sums up the squares of these variances, divides by the number for the population size, or the number minus one for population the sample size and takes the square of the result. For a normal distribution, coverage of 1, 2, or 3 standard deviations are important values with each standard deviation on either side of the mean covering, or representing depending on population or sample, increasing percentages. This coverage, six sigma for the standard deviation symbol, became a widely used term for quality control.

6.8.7 Calculus—Differential and Integral

Calculus is seen as being beyond normal fastener applications work, but I find it a useful mathematic tool to learn, and I can provide some fastener examples. Differential calculus provides solutions to rate-of-change problems. In quoting a large fastener contract to run over several years, knowing the rate of decrease in tooling costs as the order is produced gives a much more accurate true cost for the entire contract and makes the difference between a winning or losing bid and a profitable or money-losing order.

In dynamic fastener testing such as the impact strength of a shear pin, an understanding of integral calculus provides a depth of understanding in examining the stress–strain diagrams of two candidate materials and the areas bounded by their limits. Similarly, knowing the effect on volume of a change in head diameter versus a change in head height on the final upset volume can make special fastener design more efficient.

6.8.8 Differential Equations

Differential equations take the calculus and solve technical problems of a more complex but very real-world nature. Examples would be fluid flow and dynamics loading.

6.8.9 Linear and Other Programming Techniques

Some problems are encountered sufficiently often that mathematical tools and techniques can be developed to solve them. I was fortunate to take a course in linear programming where an objective function with the goal was written and the constraints written as equations. An example was a transportation problem, common in the fastener business. The goal might be minimizing transportation cost or time, and the constraints being vehicle availability or route distances. The presence of all of these mathematical tools, practiced well and ready for use, in your fastener skill set are good to have.

Reference

1. Jim Speck, P.E., *Fastener Fundamentals Seminar, National Industrial Fastener Show and Conference*, Columbus, OH, May 1994.

7

Fastener Materials

Fastener materials can proactively be specified to best meet the assembly application under consideration, and they can also be default choices based on the availability in the fastener market. By examining the many types of fastener materials possible, the assembly designer will have the best opportunity to find the right fit of material characteristics to work in his assembly.

7.1 Steels

Steel, in all its chemistries and shapes, is the most commonly used material for manufacturing fasteners. Steel is made from iron by removing iron ore's impurities in one of several types of furnaces. To have an understanding and best realize the performance potential of steel fasteners, it is valuable to understand the production processes and material characteristics of fastener steels.

Steel-making furnaces are typically very large with molten steel poured in "heats" of thousands of pounds with a heat number, which gives it traceability when combined with a mill certificate, or "cert" as they are commonly called. The mill cert with identifying heat number can give traceability to both the fastener manufacturer and user. This can give assurance that the specified fasteners are made from satisfactory raw material chemistries.

A fundamental concept is that steel is iron with the addition of small amounts of carbon and perhaps other alloying elements which increase steel's physical properties before and after processing. To visualize the effect carbon has in converting iron into steel for fasteners, recall that a diamond is the hardest naturally occurring substance known and diamond is a pure, compressed form of carbon.

Considering the metallurgical structure of any steel with respect to fastener use, it is the grains of metallurgical elements which provide engineering structure and carry the assembly loads the steel fasteners see in service. These grains of steel can be fine or coarse with a grain size rating system determined by one of several metallurgical tests. There are also an important distinction between the steel grain structure that the fastener manufacturer requires and the grain structure which is best for fastener load carrying. For the majority of fasteners which are formed with the raw steel material at room temperature, such as those which are cold headed, the grain structure is spheroidal and is achieved at the steel mill by spheroidize annealing the steel, typical in wire form and supplied in 50–200 lb coils. As its name implies, a spheroidal steel microstructure has iron, carbon and other alloying elements in a steel microstructure which has the iron and carbon, molecules in a ball-shaped grain. This rounded grain allows the steel to flow around punches and dies to form recesses, sockets, heads, head-to-shank fillets, and point chamfers. The spheroidal microstructure allows fastener forming with lower stresses. Similarly, in thread rolling, the spheroidal microstructured steel fastener blanks produce machine screw threads as well as self-threading and self-clinching features, knurls, annular and helical grooves, and radial fastener features, which have a grain flow that is continuous and unbroken. These flowed grains result in stronger metal strength in these resulting fastener features.

Similarly, in fasteners such as pins, stampings, and other formed specialty fasteners, a grain in the steel conducive to forming provides a metal fastener grain best oriented to the production of these geometries. The quality and completeness of the spheroidization can influence the manufactured fastener quality.

The grain flow of fasteners to be machined is a layered structure called laminar. In a typical section of bar stock fed into a screw machine, the grain flow orientation is parallel to the outside diameter of the raw material, as the bar is held in the collet of the screw machine and turned at the revolutions per minute which provides the correct cutting tool surface speed, these layer-like sections of laminar steel grain flow are removed by the cutting tool in the form of chips. The geometric features can be tightly held dimensionally with tolerances as small as 0.0002–0.0003 in. Sharp corners and large diameter changes are also possible. However, the interrupted nature of the grain flow means that the even after heat treatment, the static and dynamic fastening load bearing in the resulting interrupted grain flow can reduce the strength the screw machined fastener steel materials can deliver to an application. Figure 7.1 shows a comparison of cold-headed and screw-machined grain flow.

7.1.1 AISI 1010

AISI 1006 is a low-carbon steel with a nominal 0.1% carbon. AISI 1006 and 1008 are normally the least expensive fastener steels commercially available. They cannot normally be strengthened by heat treating. They do increase in strength from the wire drawing, heading, and rolling performed during fastener manufacturing, but not by a large extent.

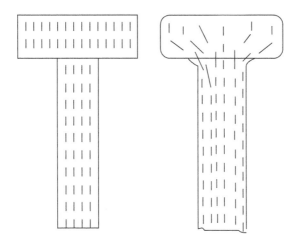

FIGURE 7.1
Grain flow of screw-machined and cold-formed fastener blanks.

7.1.2 Response to Temperature: Annealing

7.1.2.1 Typical Fastener Application

AISI 1010 and similar low-carbon fastener materials are used for machine screw, rivets, pins washers, and other low-strength fasteners.

Composition	
Carbon	0.08–0.13
Manganese	0.30–0.60
Silicon	Trace
Phosphorus	0.04 max.
Sulfur	0.05 max.
Molybdenum	Trace
Iron	Balance
Physical Properties	
Density	0.28 lb/in.^{3}
Tensile strength	60,000 psi
Yield strength	48,000 psi
Elastic modulus	30×10^6 psi
Thermal expansion	0.0000065 in./in./°F

7.1.2.2 AISI 1018

AISI 1018 and a related fastener steel, AISI 1022, with slightly higher average carbon content, are used frequently for commercial and military fasteners. It is a generally inexpensive fastener steel that can be satisfactorily case-hardened using several fastener heat-treating processes. It has good cold formability with heading and roll threading die life generally very economical.

Either case hardened or in the as cold worked condition, AISI 1018 and 1022 provide acceptable fastener mechanical strength. While AISI 1022 can sometimes be through or neutral hardened, neither it nor 1018 fasteners are well suited for this heat treatment.

Composition	
Carbon	0.15–0.20
Manganese	0.60–0.90
Phosphorus	0.04 max.
Sulfur	0.05 max.
Iron	Balance
Physical Properties	
Density	0.28 lb/in.^{3}
Tensile strength	80,000 psi
Yield strength	64,000 psi
Elastic modulus	30×10^6 psi
Thermal expansion	0.0000065 in./in./°F

7.1.3 Response to Temperature: Carbonitriding and Tempering

7.1.3.1 Typical Application

Tapping screws, self-drilling screws, blind rivet mandrels, self-clinch studs and nuts, grooved pins, drive screws, machine screws.

7.1.3.2 AISI 1038

AISI 1038 is a carbon steel having more carbon than the normal carburizing grades of 1018 and 1022. It is used for hex head cap screws as well as many higher-strength threaded fasteners. It has some of the cost advantages of the plain carbon steels. It does have lower cold-heading properties.

Physical Properties	
Density	0.28 lb/in.^{3}
Tensile strength	120,000 psi
Yield strength	102,000 psi
Elastic modulus	30×10^6 psi
Thermal expansion	0.0000065 in./in./°F
Composition	
Carbon	0.35–0.42
Manganese	0.60–0.90
Phosphorus	0.04 max.
Sulfur	0.05 max.
Iron	Balance

7.1.4 Response to Temperature: Through or Neutral Hardening

Typical application: Hex head cap screws and engineered special fasteners and parts

7.1.4.1 AISI 4037

AISI 4037 is a medium carbon alloy steel with good heat-treated tensile strength and toughness comparable to a medium-carbon steel such as 1038, while still having better cold-heading performance compared to 1038 steel.

Physical Properties	
Density	0.29 lb/in.^{3}
Tensile strength	170,000 psi
Yield strength	145,000 psi
Elastic modulus	30×10^6 psi
Thermal expansion	0.0000065 in./in./°F
Composition	
Carbon	0.35–0.40
Manganese	0.70–0.90
Phosphorus	0.04 max.
Sulfur	0.04 max.
Molybdenum	0.20–0.30
Silicon	0.20–0.35
Iron	Balance

7.1.5 Response to Temperature

Good fastener performance when through hardened from HRC 26 to HRC 43 usually is in tolerance ranges of 4 Rockwell C scale points. Tensile strength increases with hardness, but ductility declines. Set screws are through hardened to HRC 48-52 for point hardness but are brittle and used only in compression.

Typical applications: Socket product fasteners, other cap screws, and specials.

7.1.5.1 AISI 4140

AISI 4140 is a steel with fastener properties similar to 4037, but it is more highly alloyed.

Physical Properties	
Density	0.29 lb/in.^{3}
Tensile strength	180,000 psi
Yield strength	153,000 psi
Elastic modulus	30×10^6 psi
Thermal expansion	0.0000065 in./in./°F
Composition	
Carbon	0.38–0.43
Manganese	0.70–0.90
Phosphorus	0.04 max.
Sulfur	0.04 max.
Chromium	0.80–1.10
Molybdenum	0.15–0.25
Silicon	0.20–0.35
Iron	Balance

Response to temperature: Like 4037, AISI 4140 responds well to through hardening. With a more highly alloyed composition, through hardening can be carried out on larger-diameter fasteners.

Typical applications: Larger-diameter socket head and hex cap screw fasteners and larger high-strength bolting and nuts.

7.1.5.2 AISI 8640

AISI 8640 is a highly alloyed medium carbon screw, which is a premium steel for high-performance fastener products. Alloy 8740 is a similar alloy sometimes specified for high-strength fasteners.

Physical Properties	
Density	0.29 lb/in.^{3}
Tensile strength	180,000 psi
Yield strength	153,000 psi
Elastic modulus	30×10^6 psi
Thermal expansion	0.0000065 in./in./°F
Composition	
Carbon	0.38–0.43
Manganese	0.75–1.00
Phosphorus	0.04 max.
Sulfur	0.04 max.
Nickel	0.40–0.70
Chromium	0.40–0.60
Molybdenum	0.15–0.25
Iron	Balance

Response to temperature: Responds well to through hardening with both good tensile strength and toughness

Typical applications: High-performance fasteners in off-road, sports, and racing applications. Also finds application in hex keys and cold-formed fastener tooling.

7.1.5.3 AISI 52100

AISI 52100 is a ball-bearing steel with a high carbon composition. It is suitable for pins and similar fastenings, which require a high hardness.

Physical Properties	
Density	0.29 lb/in.^{3}
Hardness	Hrc 52–56
Shear strength	110,000 psi
Elastic modulus	30×10^6 psi
Thermal expansion	0.0000065 in./in./°F
Composition	
Carbon	0.95–1.10
Manganese	0.25–0.45
Phosphorus	0.25 max.
Sulfur	0.25 max.
Chromium	1.30–1.60
Silicon	0.20–0.35
Iron	Balance

Response to temperature: Responds to through hardening with hardnesses in the low- to mid-50s Rockwell C scale with high wear and shear properties.

Typical application: Pins and shear fasteners or set-screw type fasteners, which need higher hardness.

7.1.5.4 H-11

H-11 is a hot work tool steel capable of being warm and hot headed into fastener blanks, which can be subsequently thread rolled or cut and vacuum heat treated.

Physical Properties	
Density	0.31 lb/in.^{3}
Tensile strength	180,000 psi
Yield strength	153,000 psi
Elastic modulus	30×10^6 psi
Thermal expansion	0.0000065 in./in./°F
Composition	
Carbon	0.35
Molybdenum	1.50
Chromium	5.00
Vanadium	0.40
Iron	Balance less trace elements

Response to temperature: Can be vacuum heat treated satisfactorily to give a high tensile strength and good resistance to tensile strength loss at raised surface temperatures.

Typical applications: Fasteners exposed to medium temperature services of 400–700°F. An example would be screws used to hold the nozzles on plastic injection molding machines.

7.2 Stainless Steels

The idea of using metallic fasteners, which never "stain" and eliminate the process steps and expense of plating and other finishes, is an appealing one for many assembly designers. To be classified as a stainless steel, an alloy of iron must have a minimum of 11.5% chromium. A steel of an alloy chemistry containing at least 11.5% chromium will not corrode or rust when exposed to a normal atmospheric fastener service environment. "Normal" is the operative term with many service applications such as marine, sour gas, or chlorine-rich fastener service conditions presenting some stainless fastener materials with hostile environments, which may disqualify them for fastener service.

In part, this is because the industrial benchmark for a fastener material's corrosion resistance is the salt-spray test, as specified in ASTM B-117 and MIL STD 1312-1. Although the hours a fastener can last in the salt spray cabinet without exhibiting corrosion products is one measure of environmental corrosion resistance, it is based on a saline environment.

In contrast, an environment where much chlorine or sulfur is present can present a much different test outcome, especially when combined with the stress levels many fasteners operate under when in assembly.

For this reason, stainless steel of whatever chemistry should not be specified automatically as a fastener material by a fastener designer without understanding as fully as possible the hostile elements and compounds, which can comprise the intended fastener service environment. The Kesternich test, a more aggressive corrosion test, can provide useful information on fastener corrosion resistance in the kind of sulfur-rich fastener service environments found in many industrial areas. Long-term on-site evaluations can also give useful data, which can help fastener specifiers make informed material decisions.

Stainless steel is a good fastener material specification if it meets the strength, appearance, and reusability criteria you require.

7.2.1 Type 302 HQ

Type 302 HQ is a very popular 18-8 austenitic stainless steel used in the manufacture of cold-headed fasteners of many configurations. On a tonnage basis, it is by far the most heavily used stainless-steel fastener material. All 18-8 stainless steels contain nominally 18% chromium and 8% nickel. Type 302 HQ has in addition 3%–4% copper, which greatly increases its cold headability.

Physical Properties	
Density	0.29 lb/in.^{3}
Tensile strength	105,000 psi
Yield strength	75,000 psi
Elastic modulus	28×10^6 psi
Thermal expansion	0.0000096 in./in./°F
Composition	
Chromium	17.0–19.0
Nickel	8.0–10.0
Silicon	1.00 max.
Phosphorus	0.040 max.
Sulfur	0.030 max.
Manganese	2.00 max.
Copper	3.00–4.00
Iron	Balance

Thermal response: Type 302HQ can only be strengthened through work hardening. It is important to note that yield strength increases less rapidly than tensile. Full anneal to dead soft is possible.

Typical applications: Screws, bolts, pins, and many other standards and specials.

7.2.2 Type 303

Type 303 stainless steel is a bar stock austenitic material used extensively in the screw machining industries. It turns to a good surface finish when proper speed and feed rates are used.

Physical Properties	
Density	0.29 lb/in.^{3}
Tensile strength	85,000 psi
Yield strength	35,000 psi
Elastic modulus	28×10^6 psi
Thermal expansion	0.0000096 in./in./°F
Composition	
Carbon	0.15 max.
Manganese	2.00 max.
Silicon	1.00 max.
Sulfur	0.15 min.
Chromium	17.00–19.00
Nickel	8–10
Iron	Balance

Thermal response: Can be full annealed to dead soft.

Typical applications: A wide range of automatic and Swiss screw machine internally and externally threaded fasteners, nonthread fasteners, pins, spacers, standoffs, and engineered special fasteners.

7.2.3 Type 305

Type 305 is an austenitic stainless steel with a lower rate of work hardening than type 302 HQ as well as with less change in magnetic permeability when cold worked.

Physical Properties	
Density	0.29 lb/in.^{3}
Tensile strength	95,000 psi
Yield strength	54,000 psi
Elastic modulus	28×10^6 psi
Thermal expansion	0.0000096 in./in./°F
Composition	
Carbon	0.12 max.
Manganese	2.00 max.
Silicon	1.00 max.
Phosphorous	0.04 max.
Sulfur	0.03 max.
Chromium	17.00–19.00
Nickel	10.00–13.00
Iron	Balance

Thermal response: Full annealing only.

Typical applications: Fasteners requiring lower retained magnetic permeability such as those in proximity to electronically sensitive circuits as well as those requiring lower work hardening.

7.2.4 Type 410

Type 410 is a general purpose corrosion and heat-treatable fastener stainless steel. It has fair corrosion resistance and machining properties.

Physical Properties	
Density	0.28 lb/in.^{3}
Tensile strength	170,000 psi
Yield strength	136,000 psi
Elastic modulus	29×10^6 psi
Thermal expansion	0.0000056 in./in./°F
Composition	
Carbon	0.15 max.
Manganese	1.00 max.
Silicon	1.00 max.
Phosphorous	0.04 max.
Sulfur	0.030 max.
Chromium	11.5–13.50
Iron	Balance

Thermal response: It is important to note that the AISI 410 chemistry calls out a maximum carbon limit but no minimum. To vacuum heat treat 410 fasteners to nominal HRC 40 tensile, carbon should be in the upper composition range.

Typical applications: Heat-treated stainless tapping, drilling, and machine fasteners.

7.2.5 Type 430

Type 430 is a non-heat-treatable ferritic stainless steel used for its excellent cold-heading properties. On fasteners with extreme upset ratios, the low work-hardening rate allows geometries to be formed at room temperature that otherwise may not be possible.

Physical Properties	
Density	0.29 lb/in.^{3}
Tensile strength	75,000 psi
Yield strength	45,000 psi
Elastic modulus	30×10^6 psi
Thermal expansion	0.0000065 in./in./°F
Composition	
Carbon	0.12 max.
Manganese	1.25 max.
Silicon	1.00 max.
Chromium	14.00–18.00
Iron	Balance

Thermal response: Type 430 cannot be heat treated for strength but can be annealed.

Typical applications: Rivets and pins as well as special parts are typical fastener uses. Type 430 can be furnished for a high luster. It is also plated for increased corrosion resistance.

7.2.6 17-4 PH

17-4 PH is a precipitation hardening grade of stainless steel that can be used to manufacture fasteners that have high tensile strength and corrosion resistance.

Physical Properties	
Density	0.29 lb/in.^{3}
Tensile strength	180,000 psi
Yield strength	153,000 psi
Elastic modulus	30×10^6 psi
Thermal expansion	0.0000065 in./in./°F
Composition	
Carbon	0.04
Manganese	0.40
Silicon	0.50
Chromium	16.50
Nickel	4.25
Columbium	0.25
Copper	3.60
Iron	Balance

Thermal response: 17-4 PH, with nominally 17% chrome and 4% nickel, is vacuum heat treatable.

Typical applications: Machine screws, special socket set screws, and certain Mil Spec fasteners.

7.2.7 A 286

A 286 is a high-temperature fastener stainless alloy that is cold or warm headable. It has been used in the aerospace industry and has also been specified in other high-performance assemblies.

Physical Properties	
Density	0.29 lb/in.^{3}
Tensile strength	160,000 psi
Yield strength	103,000 psi
Elastic modulus	30×10^6 psi
Composition	
Carbon	0.22
Nickel	24
Chromium	14
Manganese	0.53
Silicon	0.22
Sulfur	0.002
Molybdenum	1.2

Thermal response: A286 work hardens during heading and thread rolling. Warm heading increases formability.

Typical applications: Typical A286 fastener applications would be military and aerospace assemblies where some mildly elevated service temperatures are inherent.

7.3 Nonferrous Materials

Both steel and stainless steel fastener materials are ferrous in that they contain iron as part of their chemistry. Many fasteners are made from nonferrous metals. These materials contain no iron in their makeup. Some popular and different nonferrous fastener materials are brass, bronze, aluminum, and titanium. Brass is an alloy of copper and zinc. Copper is the primary element, composing up to 70% in some brass alloys. The higher the copper content, the more reddish is the color the resulting fastener turns out. With less copper, a yellow fastener color results. Of course, the color is directly related to after-fastener manufacturing processes such as washing and burnishing. If not protected with some fastener finish such as a plating or lacquer, oxidation can quickly tarnish the brass fastener finish. Just be aware that just as with any finish, lacquer or plating usually alters the thread and bearing surface coefficient of friction and resulting clamp load at a given torque applied to threaded fasteners. Alloy 260 (60% copper, 40% zinc) and "cartridge brass" are two yellow brass material examples.

Aluminum is mined from bauxite ore and is used to provide a lightweight fastener material combined with some fastener strength. With a density almost one-third of their steel counterparts, aluminum fasteners offer significant weight savings in assembly. With anodizing, attractive and protective finishes can be applied.

7.3.1 Spectrographic Analysis

Often, it is desirable to check the chemistry of a fastener after it has been manufactured and delivered, perhaps with the traceability back to the raw material heat having been lost. The need for a chemistry check can be for auditing purposes, to assure that the fastener material ordered has been delivered. Fastener chemistry verification could also be required for purposes of troubleshooting an application's fasteners or doing failure analysis work. A useful piece of laboratory equipment used for fastener chemistry evaluation and verification is the spectrographic analyzer. It is found in many commercial, OEM, and manufacturing laboratories.

In use, a section of fastener raw or finished material is sectioned and flattened to fit in the spectrographic chamber. An arc is struck against the sample by the unit in a manner similar to an arc welding rod. The light given off by this arc, when it is performed in an inert atmosphere, travels along a wave guide to a reflective grid, where it strikes at an angle. The light is then split, with each element's light bouncing off the grid at its own specific angle. Light-sensitive collectors, which are arranged in an arc at each element's angle, absorb the light coming in, convert it to electrical potential, and send it to a display circuit when prompted by the machine program or operating technician. In this manner, a spectrographic analysis can identify fastener composition within a short time period.

7.3.2 Carbon Analysis

Another piece of lab equipment used in chemistry verification and checking is a carbon analyzer. With this apparatus, the other elements in a sample specimen are separated out and the resulting carbon weighed for verification.

7.3.3 Magnetic Screening

Magnetic screening equipment uses the varying magnetic field response of different fastener microstructures to differentiate fasteners. If batches of fasteners are to be screened for like chemistries or heat treatment, their magnetic field response can be used to identify similar and dissimilar fastener structures.

7.3.4 Eddy Current Techniques

An inspection device based on the above magnetic response phenomenon is the eddy current sorting machine, which uses pinball machine style flippers driven by the output of magnetic coils as fastener samples pass through at relatively high speed. They can be used to cull out different fastener metallic structures in lots with mixed structures. Sound, laser, optical, and mechanical inspection machines can also be used for sorting fastener lots for other features. Some trained technicians can also detect material differences by the shape and color of grinding "roster tails" from fasteners held to an abrasive wheel. Figure 7.2 shows a sketch of the spectrographic chemistry analysis.

7.3.5 Brass Alloys

Several brass alloys are used for screw-machined, cold-formed, and stamped metal fasteners. They are similar in physical properties, physical composition, and service environment response.

	Alloy 260	Alloy 270	Alloy 230
Copper (%)	70	65	85
Zinc (%)	30	35	15
Tensile strength (psi)	90,000	74,000	61,000
Yield strength (psi)	67,500	55,500	47,750
Elastic modulus (psi)			
Density	0.30	0.30	0.30

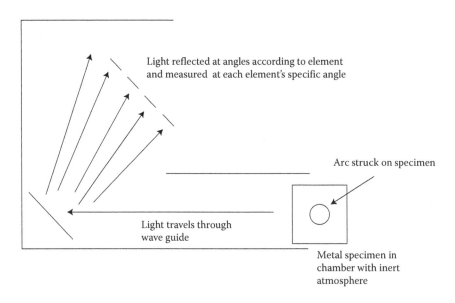

Light reflected at angles according to element and measured at each element's specific angle

Arc struck on specimen

Light travels through wave guide

Metal specimen in chamber with inert atmosphere

FIGURE 7.2
Sketch of spectrographic analyzer.

Thermal responses: Cannot be strengthened by heat treatment, only by cold working. Brass fasteners such as rivets or eyelets can be annealed.

Typical applications: Various screws, rivets, and related fasteners and hardware.

7.3.6 Aluminum Alloys

Aluminum alloys are used for fasteners where light weight is an important assembly service criterion. Several alloys are commonly manufactured into fasteners, with individual aluminum alloy selection dependent on strength requirements.

	1100	2024	6062
Composition	Al = 100%	AL, Cu, Mg	Al, Cu, Cr, Mg
Tensile strength (ksi)	13,17,24	26,64,64	16,30,37
Yield strength (ksi)	5,16,22	11,44,42	8,20,32
Density (lb/in.²)	0.11	0.11	0.11
Tempers	0,H14,H18	0,T3,T4	0,T6,T42
Thermal expansion	0.0000126 in./in./°F		

Thermal response: Fasteners made from aluminum can be heat treated by most commercial heat-treating shops to the desired temper designations. Some of the higher-numbered tempers can involve significant time to age to strengthen.

Typical applications: Screws, rivets, blind rivet bodies, and mandrels.

7.4 Other Fastener Materials

7.4.1 Titanium

Titanium is a low-density, high-strength fastener material. It is used for aerospace and demanding applications where the fastener's high strength-to-weight ratio is an assembly requirement. It is not easily formed into fasteners. That, combined with a high raw material cost, makes it a premium fastener material.

Physical Properties	Composition	
Tensile strength	Ti	90%
Yield strength	Al	6%
Density	V	4%

Typical Applications: Aerospace and sports assemblies where high strength with low weight are required and the higher cost of fastener manufacturing in titanium is not prohibitive.

7.4.2 MP35N

MP35N is a proprietary fastener material of SPS Technologies, Jenkintown, PA and produced by Carpenter Technology Corporation.

Physical Properties		Composition	
Tensile strength	227,000 psi	Ni	33/37%
Yield strength	217,000 psi	Cr	21/19%
Density	0.304 lb/in.^{3}	Mo	9/10.5%

7.4.3 Monel

Monel is a fastener material often identified with marine applications.

Physical Properties		Composition
Tensile strength	140,000 psi	Nickel 65%
Yield strength	100,000 psi	Copper 29.5%
Density	0.306 lb/in.^{2}	Iron 1%

7.4.4 Hastelloy

Hastelloy is an aerospace fastener alloy with good high-temperature creep resistance. This makes it useful for applications where high temperatures are part of the service environment. An example would be jet engine turbines.

Physical Properties		Composition
Tensile strength	110,000 psi	Nickel 57%
Yield strength	47,000 psi	Molybdenum 20%
Density	0.318 lb/in.^{2}	Iron 20%

7.4.5 AerMet 100

AerMet 100 is a proprietary material of Carpenter Technology Corporation used for high-performance fasteners. One application is for a very-high-strength bolting. It combines high hardness and tensile strength with excellent ductility and toughness.

Physical Properties		Composition	
Tensile strength	285,000 psi	Ni	11.1%
Yield strength	250,000 psi	Cr	3.1%
Density	0.285 lb/in.^{3}	Co	13.4%
		Fe	Balance

Typical applications: AerMet 100 alloy is a candidate fastener material for aircraft, ordnance, high-performance automotive and "ballistic tolerant components."

Effects of Alloying Elements in Fastener Steels
These alloying elements generally add properties to steel fastener materials as described for the element, listed by normal significance with respect to fasteners. Each element's periodic chart descriptor is in parentheses.

Carbon (C): Carbon is one of the most important elements in, or rather constituent of, steel. It increases a fastener's tensile strength, hardness, and resistance to wear and abrasion. It does, however, also lower fastener ductility, machinability, and toughness.

Manganese (Mn): Manganese is a deoxidizer and degasifier. It also reacts with sulfur to improve forgeability. Manganese increases tensile strength, hardness, hardenability, resistance to wear, and the rate of carbon penetration in carburizing. It also decreases the tendency toward scaling and distortion.

Molybdenum (Mo): Molybdenum increases strength, toughness, hardness, and hardenability as well as creep resistance and strength at elevated temperatures. It improves machinability, corrosion resistance, and intensifies the effects of the other alloying elements. It also increases hardness at raised temperatures.

Aluminum (Al): Aluminum is a deoxidizer and degasifier. It retards grain growth and is used to control austenitic grain size.

Chromium (Cr): Chromium increases tensile strength, toughness, hardness, and hardenability, as well as resistance to wear and abrasion. It also increases resistance to corrosion and fastener scaling at raised temperatures.

Nickel (Ni): Nickel increases strength and hardness with no loss of ductility or toughness. It also increases resistance to corrosion and scaling at raised temperatures when used in suitable quantities in high-chromium stainless steels.

Sulfur (S): Sulfur improves machinability in free machining steels. It decreases fastener ductility and impact strength. Also, the addition of sulfur without sufficient manganese produces brittleness at red heat.

Silicon (Si): Silicon is a deoxidizer and degasifier. It also increases the tensile and yield strength, formability, hardness, and magnetic permeability. It can reduce fastener cold-heading, machining, and roll threading tool life.

Copper (Cu): Copper improves fastener resistance to atmospheric corrosion and increases the tensile and yield strength with a reduced loss in ductility.

Vanadium (V): Vanadium increases strength, hardness, and impact resistance. By retarding grain growth, vanadium permits higher quenching temperatures. It also improves the red hardness properties of fasteners made of tool steels such as H-11 and intensifies the individual effects of the other major elements.

Boron (B): Boron is a potent and economical addition to a fully deoxidized steel, normally used in alloy steels. Added to the heat in very small amounts of the order of 0.01%, it can dramatically increase hardenability. Boron cannot be added in large amounts without causing raised temperature brittleness.

Cobalt (Co): Cobalt increases strength and hardness in addition to permitting higher quenching temperatures. Also, it intensifies the effects of the other major elements in more alloyed fastener steels.

Nitrogen (N): Nitrogen is a strong austenitizer, which can substitute for a portion of the nickel in stainless steels. It is also used as an element in carbonitriding many fasteners containing aluminum or chromium to produce an extremely hard case. It is added to the heat of some steel alloys to enhance machinability by the production of very fine chips when the fastener is turned, slotted, milled, or machined.

Tantalum (Ta): Tantalum is used as a stabilizing element in stainless steels. It has an affinity for carbon and forms carbides, which are evenly dispersed throughout the steel, thus preventing localized depletion of carbon at grain boundaries.

Titanium (Ti): Titanium is added to stainless steels to make them resistant to harmful carbide precipitation.

Tellurium (Te): Related to sulfur and selenium in the chemical classification of elements, tellurium has the similar effect of improving machinability when added in small amounts to some free machining steels.

Columbium (Cb): Columbium in stainless steel has an effect similar to titanium and tantalum in making the steel more resistant to carbide precipitation and the resulting intergranular corrosion.

Tungsten (W): Tungsten increases strength, toughness, and hardness. At elevated temperatures, tungsten steels have superior raised temperature characteristics.

Lead (Pb): While not strictly an alloying element, lead is added to some steels to improve machinability. It is almost completely insoluble in steel and minute lead particles dispersed throughout the steel reduce friction where cutting tools contact the fastener work piece. It also improves chip breaking during turning.

7.4.6 Notes on Fastener Metallurgy

There are two things to keep in mind in reviewing a fastener's metallurgical composition and drawing conclusions concerning fastener performance. First, each element can contribute to the fastener's various strengths, such as tensile, impact, or creep, either independently or in concert with the other alloying elements. This happens through the microstructure, or grains of fastener metal, and the influences of working during manufacturing and thermal responses during any heat treating that is undertaken.

Secondly, the percentages of each element can, and do, vary with each batch or "heat" of fastener raw material. Variances in fastener metallurgy within specification can affect fastener performance. If a specific fastener performance characteristic is critical, understand and control it to the extent possible. It is also good practice to ensure compliance to chemistry specification through ongoing monitoring and periodic spot checking through accredited laboratories set up to perform alloy identification and verification.

7.4.7 Brazing Alloys

Metal filler metals used for brazing have lower melting temperatures than the assembly materials being joined. An approximate braze melting temperature threshold would be 1200–1600°F. Some assembled member materials are aluminum and aluminum alloys, copper alloys, ferrous metals, magnesium alloys, and 300 as well as 400 series stainless-steel alloys.

Some brazing elements and their percentages are as follows:

1. Gold 10%–80%
2. Copper 16%–93%
3. Zinc 4%–65%

The strength of the brazed joint assembly and its mechanical fastening performance can be varied by the selection of brazing alloy percentages.

7.4.8 Nonmetallic Fastener Materials

Solid fasteners—A growing demand for plastic fasteners has resulted in plastic fasteners of two general types. The first are screws, nuts, washers, and rivets, which are plastic versions of their metallic counterparts, for example, pan or hex head machine screws. The second is innovative design for assembly of plastic fasteners, which are efficient in use and attractive in appearance.

The most common plastic used for solid plastic fasteners is Type 6/6 Nylon. It can provide some electrical insulating properties, which differentiate it from metal fastener materials. This is a characteristic which can be useful in fastening computer equipment, business machines, and other electrically powered assemblies. Nylon 6/6 is tough, resistant to some of the corrosion that can rust plated steel fastenings, and light in weight. Appearance is a positive material property with many plastic fasteners. Nylon is hygroscopic, meaning it changes dimension as it absorbs water. From the mold, plastic fasteners contain approximately 0.05% moisture and are a natural color. After dyeing to add color, the nylon content can increase to 3%–4%. Low moisture content results in more rigidity. As moisture content increases, so does impact strength and toughness.

7.5 Fastener Materials in Today's Environment

One of the most important developments in fastener materials since the writing of the first edition is the introduction of country of origin control of fastener raw materials used in fasteners for MS, AN, and NAS fasteners. Where a fastener raw material comes from has become an issue for the fastener industry. Based on previous experience, this in itself should not come unexpectedly. The strength of a mechanical fastening is dependent on its material strength properties as well as its dimensional features. The integrity of the properties can be compromised though inadvertent or intentional variance in chemical makeup. Quality systems can play an important role in fastener metallurgy. A cornerstone of an efficient and effective quality system is a three-step process:

Say what you are going to do

Do what you say

Document that you did

What will we do in our fastener laboratory? To answer that question, let us consider some of the possible equipment one would expect to find. At the start, dimensional gages, including micrometers, calipers, and scales would be fundamental. These will enable us to measure fastener diameters and lengths. We could have inch and metric unit micrometers and calipers or digital units that can be toggled to measure in either system. With respect to accuracy, our micrometer will read more finite increments than our calipers. Calipers will work for quick measurements and micrometers for more

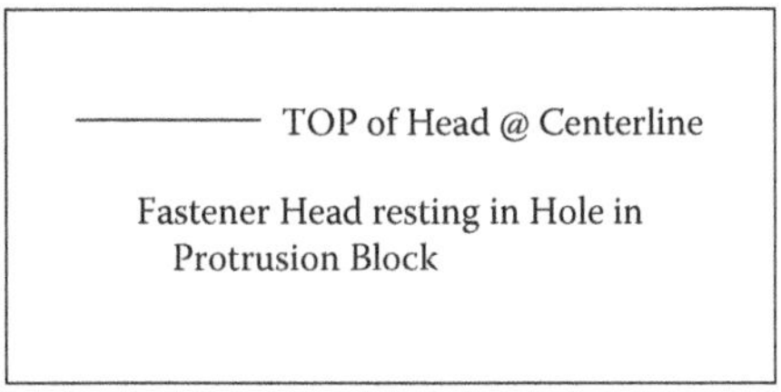

──────── TOP of Head @ Centerline

Fastener Head resting in Hole in
Protrusion Block

FIGURE 7.3
Measuring head protrusion.

precise measurements. Remembering the laboratories where I started out in the fastener business, the labs were separated into dimensional and metallurgical departments. Later in my career at a smaller fastener manufacturer, these functions were combined. Continuing with dimensional equipment, it would be useful to have blocks to measure the protrusion of 82°, 90°, and 100° countersunk heads. Ninety-degree flat-head screws are a standard in metric standard screws while 82° and 100° included angle heads are both inch standard designs. A protrusion block is shown in Figure 7.3.

Measuring head heights is often done best by measuring over the centerline axis of the screw with the aid of a protrusion block. On flat head screws, it allows any inspector to accurately obtain the protrusion height and avoids the technical issue of locating where on the head-to-shank fillet radius the head starts. It also provides an efficient underhead datum for most other common head styles. It can often be aided by cutting off any fastener shank section longer than the block providing an inspection sample can be scrapped without economic difficulty.

An optical comparator is usually to be found in the fastener laboratory. With appropriate magnification lenses such as 10×, 50×, and 100× and optical comparators and clear plastic overlays, angles, radii, and other fastener geometries can be inspected. Important features of a satisfactory optical comparator will include a provision to swivel the table upon which the samples are stages through a sufficient number of rotational degrees to align screw thread roots for their specific lead angles as well as a digital X–Y reader to measure the distance traversed. With these features, fastener features such as pitch-to-pitch lead can be inspected.

This simplistic outline encompasses quality plans and procedure, training, records, equipment, and calibration, and many of the other steps in an organization's quality manual. With respect to fastener materials, that would include requiring raw material certificates of the raw material's chemical and physical properties, lot control, as well as raw material, and final fastener product testing.

Putting it all together: Some thought on screw threads engaged and under load in service:

As stated previously, screw threads give threaded fasteners the user benefit of reusability, which along with strength, economic efficiency, and appearance constitute important factors in their use. Consider an external thread of some length, say eight times its diameter, which we will call 8D. It is to be threaded into an internal thread that is one diameter in thickness, which we will call 1D. We know that standardized metric and inch system screw threads, as well as special fastener screw threads, have specifications for major, pitch, and minor diameters as well as pitch, lead, helix, and flank angles. To be able to meet the required screw thread specification, each of these dimensions on our external and internal screw threads must be within specification tolerances. Within these tolerances, however, variations have an effect on the performance of the threaded, fastened assembly. It is important to the understanding of fastener performance to study and understand the effects of threaded and fastener-related geometry variance, within tolerance, on the static and dynamic performance of the resulting threaded fastener assembly, under load conditions, in service. The general term we will use to examine these variances in fastener geometry is run out. Cylindrical and perpendicular runout is the dimensional departure from the center or perpendicular axis of a fastener's geometric features. A method to communicate fastener runout is to study a pair of externally and internally threaded fasteners, each at maximum material condition and with zero runout. If the reader will recall, maximum material condition is the dimensional limit of a component where the external part could not be any larger without being oversized to its specifications and the internal part could not be of a tighterfit without being outside of its specifications. On metric threaded fasteners, class 6h screw threads engaged with 6H screw threads, each at maximum material conditions would be a line-to-line fit. There would be no looseness, or "play" between the engaged screw threads. Similarly, inch system 3A and 3B screw threads would be of this line-to-line tightness of fit. Consider our threaded components in three-dimensional space, we can inspect both external and internal threaded parts for variance from this perfect fit. As each part moves away from its maximum material condition, the allowance or room varies from a true position with respect to its X-, Y-, and Z-axes. In aircraft terms, each can have roll, pitch, and yawl with respect to its assembly position with its mating part. And it is very important in the understanding of mechanical assembly performance that these variances, or runouts, have an effect on the static and mechanical performance of these fastened assemblies. As an example, industry standard perpendicularity of bearing surface square-ness allows a 2° variance from perpendicular. In side view, a bolt head can be cocked up to 2° and be within specification. However, when this part is tightening into a mating part with a perfectly square mating surface, it will, if it is able, square back up. In doing so, some amount of bending stress in this bearing area is a resultant. Stresses induced in the installed threaded fastener can reduce the total available fastened performance and capability of the assembly affected by compounds with excessive amounts of runout. If we compare the wedge tensile and fatigue test results for two threaded assemblies of greatly different runout, we would expect to see significant test result data.

7.6 The Fastener Laboratory

Testing is a fundamental part of fastener application engineering. For many fastener practitioners, the ability to measure, test, and try fasteners and fastener assemblies is an important step toward better assembly performance. The fastener laboratory can often be the best location to perform these tasks. Fastener lab equipment can range for a work bench and some simple hand tools to multimillion dollar complexes with state-of-the-art equipment for metallurgical, chemical, and physical testing and analysis.

Absolute versus relative accuracy is a distinction that I find useful. The laboratory is concerned with absolute accuracy to known standards. The gage user is often most concerned with the relative measure value of one fastener to the lot history, or "signature." The difference between precision and accuracy is that the former refers to the repeatability of the operation, whereas the latter refers to the closeness of the indicated value to the true value. In making any type of fastener laboratory measurements, the lab technician must properly be concerned with the truth of the measurements being reported. Two important concepts in laboratory measurements are the chain of traceability and the uncertainty of any measurement. To know that 10 cm is truly 10 cm and 1 kg is exactly 1 kg, we compare our measurement equipment, by calibrating it, to a known reference. This reference artifact is then compared to another known standard. Variances, or measured differences must be smaller than acceptable laboratory standards. This chain of comparisons should be continued through a traceable path. Most countries have in place an operating network for all of the equipment we might find in our fastener laboratory. This list of measuring tools and equipment (MT and E) will usually include identifying numbers, calibration dates and frequency, external calibration sources, and other pertinent data. In the fastener laboratory, it is not uncommon to make a fastener inspection report, which will later be used by customers as a basis for installation decisions. In the event of a fastener and/or assembly failure, the information on the test report is only as good as the fastener laboratory equipment used to make the measurement or complete the test. For this reason, the traceability of the measuring or testing equipment's calibration provides an all-important understanding of its accuracy. Let us, at this point, define accuracy. A good working definition for fastener laboratory use is, "How close is the information to the truth." Taking screw thread pitch diameter as an example, accuracy of a pitch diameter measurement would be how close it is to the actual pitch diameter. Many factors can influence even a slight variance from the pitch diameter that an actual fastener's screw thread will provide in the assembly as it is being installed. Some are equipment variances such as gage manufacturing tolerances, wear, temperature, and pressure changes along with many other equipment variances. Other variances can be from the inspector and parts such as technique used, location on the fastener's thread length, and similar factors that can vary slightly. As an example of variance in accuracy, one specific fastener's screw thread could have its pitch diameter measured in different fastener laboratories and with different equipment and have slightly different values reported. For the fastener person, the logical question becomes, which is true? The inspection equipment's accuracy guides informs users in answering this question. If a measurement is well centered in a tolerance, these small variances may not be significant. If, however, the measurement is on a borderline limit of being within, or outside of its specification limit, this data and its accuracy are very important. The occurrence of borderline inspection date occurs much more frequently than one would expect in fastener applications. Note that accuracy differs from precision of measurement. A fastener lab device can have precision in that its reading on a specific inspection repeats to a very close variance. That repeating value, however, can vary from the true value, just repeatedly. A qualified piece of fastener laboratory equipment should be both accurate and precise. Both accuracy and precision are important. And knowing the difference is a key concept for the fastener professional.

An idea and technique that ties this together is uncertainty. Uncertainty is a statistical term. For our fastener laboratory review, we can introduce the term gage uncertainty. Gage uncertainty provides a measure of the distance that any quality fastener measurement varies from the true value. In this respect, it is a measure of the laboratory equipment's accuracy. The uncertainty of any device used in fastener work is composed of two major parts. These are the device itself and the device's operator and environment. The factors that comprise the variance of the device itself are in large part dependent on the design of the equipment or gage and the skill and quality of its manufacture and maintenance. The gage will have tolerances for its manufactured parts. The gage may be in accordance to some industry standard. A rule of thumb is that the variance should not exceed 10% of the product tolerance. If the tolerance, for example, for cylindrical runout of an unthreaded body of a component to its pitch diameter cylinder is 0.2 mm, then the variance of the gage should be within 0.02 mm. The gage should have repeatability within this range. The federal Standard H28, which is an important North-American document on screw threads for federal service, recommends the best method for checking an indicating thread gage is that it repeat on its respective master setting gage. The measurement operator, the fastener being inspected, and its environment comprise the second component of gage uncertainty. Operators can affect the reading given by a specific and often unconscious technique, which has a specific bias. An example would be heavier than average hand pressure while holding the fastener being inspected. In the author's experience, thread ring gages in particular are prone to this type of operator-induced effect. Variation in the fastener feature being inspected by the lab equipment will by its varying geometry, even within part tolerance, provide the operator varying readings on the same part with a very accurately repeating gage depending on the location on the part of each measurement. And as the gage is used, it may be subject to some amount of wear, perhaps at its point of contact with successive fasteners being inspected. With adequate recalibration, this variation could make the certainty of the fastener inspection result not valid. Even when gage wear is well calibrated and within its gage tolerance, readings made later in a calibration cycle can have significant variance and, therefore, have great uncertainty. The understanding of the gage uncertainty of each fastener inspection can prove to be useful information for both fastener engineers and technician, and fastener company managements. Developing gage uncertainty measurement can be performed in one of two standardized methods. These are detailed in Guide to Uncertainty Measurement industry standards, or GUMs. The first entails creating an uncertainty budget using factors of interest for the equipment and inspection environment, including those mentioned previously. It is a rigorous statistical study and worthy of review. The second method is more heuristic and is based on the informed estimate of knowledgeable evaluators experienced with the equipment. Where this can prove useful is cases where a fastener measurement is on a specification limit and may be in dispute by other labs, inspection operations, or users of the fastener data. As an example of using uncertainty, the author was involved in the shear testing of an aerospace rivet manufacturer manufacturing to a military standard. The test procedure was well specified in the procurement document called out in the part specification. The test was performed on a tensile tester set up for the double shear test required, and the required sample number tested all met the specification. The end user contracted a well-established commercial laboratory to replicate the test and the sample tested just below the specification's lower limit. Review of both tests showed little scatter in the test data and the author's and commercial laboratory's results straddled the specification. Retests by both showed similar test results. As a step to get to the root cause, the calibration records for both tensile testers were reviewed. The test standard required no greater than 1% variance. While both pieces of lab equipment meet the standard, the commercial laboratories variance was significantly great. Based on this review, the parts were accepted.

Another widely used numerical technique for accessing the accuracy of fastener measuring and testing equipment is the Gage Reproducibility and Repeatability Test, commonly known as the Gage R & R. In this technique, samples of 10 pieces are inspected twice or thrice by two to three operators. A sequence of calculations are performed with the result, providing a measure of how the gage, operator, and sample repeat for each inspector. Measuring each sample multiple times by multiple individuals gives a quantitative indicator of measurement stability. In this case, reproducibility denotes different operators measuring the same sample with the same device. The total Gage R & R value is typically given as a percentage of the fastener feature being inspected for tolerance. This can often be part of a capability of a preproduction approval process, or PPAP. Typical acceptable Gage R & R percentages are in the mid-teens, with 15% being widely used. The author has seen data for below 5% as well as much higher. A higher Gage R & R number should be studied for the inspection process as a system, including part and operator variance before determining that a particular fastener inspection device is not capable. Large swings in the percentage are possible with relatively small changes.

Knowing the uncertainty of any fastener laboratory MT and E is a fundamental for professional fastener testing and inspection.

Reviewing additional equipment, we can expect to find in a laboratory set up for fastener work would include devices for dimensional measurement, mechanical testing, metallurgical analysis, performance testing, and resistance to environmental conditions. We started with micrometers and calipers, and optical comparators, for dimensional inspection of fastener features. Gage blocks, inspection fixtures, and recess penetration gages might be found. Surface roughness devices, microscopes, a surface plate, and mold-making compound would also be of use. For the dimensional inspection of threads, limit gages would be comprised of thread ring and plug gages. The availability of qualified master setting plug gages for thread go and no go ring gages is advisable as ring gages can wear out of tolerance, usually class X for work rings and plugs, and class W for thread masters. A use between calibrations for attribute gages of 200 uses can occur very quickly in a busy fastener inspection lab. With more accurate screw thread quality information are variable thread gages. These would be available for internal and external screw threads. They would have masters for setting and checking the indicating gages, as well as analog or digital indicating dials or readout. The gage bodies and frames would be mounted with rolls or segments for measuring pitch diameter and functional size, analogous to the attribute gage's go and not go members, but in actuality it measures the fastener screw thread's minimum and maximum material conditions. The differential between pitch diameter and functional diameter of the inspected fastener can be thought of as a measure of its thread form variance from ideal form with respect to angles, lead, and circularity.

In the metallurgical laboratory, it would be typical to find standard, superficial, and microhardness testers; spectrographs; metallurgical mold-preparation equipment; plates for driving self-threading screws; and torque wrenches. A good tensile tester with compression capability and test fixtures for shear testing of fasteners is often found being used frequently. Fatigue and impact testers as well as creep and hot tensile testing equipment would be more specialized. For testing of cap screws, ASTM wedges for wedge tensile testing might be found. For the detection of surface imperfections, liquid penetrant and magnetic particle testing equipment may be required. In many cases, these laboratory tests may be subcontracted. For corrosion-resistant fasteners, the author had obtained and used a magnetic permeability tester.

For the detection of surface iron residue remaining after fastener cleaning from process tooling, a copper sulfate test station proved useful.

In the environmental testing area, a salt spray tester and humidity cabinet can prove useful for testing fastener and fastened assembly resistance to corrosion. For evaluating fastener plating thickness, plating thickness test equipment is available for the lab. As some fastener finishes are subjected to more sulfur-rich, as opposed to salt-rich, marine, environments, apparatus is available to provide these environmental conditions in the laboratory. Specialized equipment for hydrogen embrittlement, many dynamically loading tests such as vibration, impact, and crash application can also be found to verify and improve these fastened assembly factors. I am reminded of a piece of advice I was given early on with regard to fastener testing. It was that when testing using lab equipment and procedures for a specific purpose, it will sometimes occur that a seemingly unrelated phenomenon will catch the inspector's attention. It is of value to note these observances for future investigation. These follow-ups can greatly increase one's fastener skills and understanding and lead to profitable new products.

It is not only individual fasteners that can be inspected. Often, it is useful to inspect and test complete assemblies. Some examples that have been observed use the previously mentioned drive plates and torque wrenches to perform drive torque and stripping strength tests of assemblies fastened with self-threading fasteners. Early in my career in the fastener industry, pull tests were performed on street sign stainless steel strapping fastened with stamped stainless-steel buckets tightened with my employer's stainless-steel cup point socket set screws. The tests were performed on a tensile tester fitted with the appropriate grips. As a baseline, strapping by itself was pulled to failure to obtain its yield, ultimate tensile, and break strengths. Next, completed assemblies were assembled with the set screws properly torqued to specification. By pulling these to failure, an index of performance could be provided to the customer to demonstrate joint efficiency. I can still picture an early carbon fiber aerospace component being fatigue tested with the amount of cooling air provided by a fan and varying orientation of the carbon fiber to study the effects of these factors on the assembly's fatigue performance. With the use of strain gages, the decreases in tensile fastener preload and loading over time have been successfully plotted. These are a few of the many fastened assembly tests that a skilled and focused technician and a good laboratory can study, understand, and improve. The following decision flow chart can provide a useful set of steps in the laboratory:

Test decision flow chart

Determine assembly characteristics to be tested.

Prepare applicable test methods and procedures.

Determine test measurement and test equipment required.

Decide on the number of assemblies to be tested.

Complete tests, recording data with care and professionalism.

Note corrective actions indicated if nonconformance to test parameters occur.

Data retrieved, given appropriate oversight and review.

Data given formal recording and archival, with any test specimens required.

Provide appropriate future access to the data report and supporting documentation.

7.7 Electrical Theory and Fastening, Joining, and Assembly

Electrical engineering and fastener technology do not often get connected. I completed 4 years as an apprentice electrician before matriculating from evening to full-time student to earn a bachelor's degree. I have said many times, the lessons I learned as an apprentice helped me so many times in my fastening career toward accomplishing specific fastening project goals. In this section on materials and fastening laboratories, a review of electrical and electronic engineering fundamentals will provide the reader with useful technical foundations.

Recall that all materials are composed, at the atomic particle level, of atoms whose basic structure has a nucleus of positively charged protons and neutrally charged neutrons orbited by negatively charged electrons. The difference in electrical potential between positively and negatively charged particles is a good starting point in discussing electrical theory. The electrical difference in potential between negative and positive electrical charges results in a force between them. This force can be measured, understood, and employed to do useful fastener manufacturing, testing, and installation work.

Electrons flow as a result of these electrical charge forces. A tour through a fastener manufacturing factory will show many pieces of production equipment being powered by electrical motors and supplied with electric light fixtures. The flow of electricity for these devices is measured in amperes, or amps for short. Let us take a look at how it is generated. Both the opposite charges flow and create a field force around the flow path. This field is known to us all as magnetism and, like the flow measured in amps, is measured by its magnetic field strength. Moving a good conductor such as a copper wire through a sufficiently strong field results in a current flow in the conductor. Depending on the direction of the conductor's motion, it can be direct, always in one direction, or it can alternate, first in one direction then reversing to the opposite direction. These are direct current (DC) and alternating current (AC). As an apprentice I was able to

participate in the winding of many direct and alternating current generators and motors. Let me share that with you now to help you understand
the underlying electrical theory. Electricity is generated by the movement
of a rotating conductor through a magnetic field; the field induces a flow
of electrical current, which is measured in amperes and whose potential is
measured in volts. In engineering practice, electrical current is denoted by
the letter *I* and electrical potential by the letter *E*. The field is often stationary and is called the stator. The rotating center section, called the rotor, is
typically manufactured from a specific pattern of copper wire in the form
of windings, or turns. The number of turns and their geometry, as well
as the geometry and field strength of the stator windings and pole pieces
are design elements of this electricity-generating device. For direct current,
this device is called a generator. For alternating current, it is called an alternator. It is of interest that by changing the design and having the stator's
magnetic field induce motion in the rotor instead of inducing current flow,
the device becomes an electric motor, either direct or alternating current
depending on design. For our electric generator, we will require a rotational force to turn the rotor. We call this force the prime mover. It could be
water powered as in a hydroelectric dam, it could be an internal combustion engine, or it could also be a steam turbine. Once running, the generator or alternator requires a method to transfer the electric current from the
generator or alternator. For the generator, it has a commutator which is a
series of insulated bars, usually made of copper, each one attached to a set
of rotor windings. As the generator rotates through the magnetic field created by the stator, the magnetic field induces a current flow in the rotor's
conductors as they travel through the magnetic field. A good piece of information for the fastener person is how this current is collected from the
rotor windings and conducted to a power line for performing useful work
in a factory or home. This involves fastening. In a generator, the windings
are often connected by soldered joints to the commutator bars. Brushes,
usually made from graphite in my apprentice experience, conduct this electrical current to wires as the generator's electrical output. The mathematics
underlying this flow could be shown graphically as an S-shaped or sinusoidal wave with voltage and current being the *Y*- and *X*-axes. This periodic
rise and decline of the voltage in the curve is due to the increase, peak,
and then decrease in the current flow as the rotor winding passes through
each stator pole piece. Electrical engineering design provides equations for
the number of stator poles and their magnetic field strength as well as the
rotors, diameter, and number of windings, rotational speed, and the number of commutator bars. An electrical generator output is in direct current.
Most electrical power used in the fastener as well as in other industries
is alternating current. By replacing the generator's commutator bars fastened to rotor winding, an alternator collects the induced magnetic current with slip rings and brushes. Not interrupting the current flow as the
side-by-side insulated commuter bars are, much more of the full sine wave

is collected and conducted through the brushes as alternator electrical output. In this manner, alternators are more electrically efficient in converting a larger percentage of the mechanical energy that turns the electrical device's shaft into electrical power. So far, we have introduced voltage and current. Electrical power is the ability to perform work and is expressed as current multiplied by voltage. The equation is $P = E*I$. The units of electrical output are the watt and kilowatt. In terms of time, kilowatt hour provides a measure of the ability to perform work such as power and fastener cold headers or roll threaders. Returning to our rotating electrical devices, the magnetic field of the stator windings can be used to provide rotation of the rotor by the electromotive force of the stator field. Electric motors can be either of a direct current–type, which is common for many smaller electrical motors, or of an alternating current–type, which will be found through fastener manufacturing machine tools. The unit of power is the kilowatt, although horsepower, a very old but very widely used unit of measure, is still widely used. One horsepower equals 0.7457 kW. An AC induction motor of 3.8 kW motor was used in many of the single-die two-blow headers capable of upsetting up to 5-mm diameter low-carbon steel wire in my fastener manufacturing experience. Let us turn our attention to electric power and some technology for understanding it better with mathematical tools to analyze electrical circuits. Starting at its generating source, which now could include wind-powered turbine generators and fields of solar panel arrays, the needed power flows along transmission. The humming of the wires is an example of the alternating rise and fall of the magnetic fields. The frequency of these oscillations is usually 60 times a second, also knows as 60 cycles or 60 Hertz in metric units. Other frequencies can be used. During my apprenticeship, I observed 400 cycles being used for powering electronic equipment. In the major transmission towers, these line can be 14,400 V and three phase or have three sine waves being outputted from the generating windings, with a "Y and a Delta being two three cornered winding apex winding configurations. Earlier, in the section dealing with dynamics, we spoke of forces, actions, and friction. Even in electrical theory, we have a type of systems loss, and we can simplistically identify it as resistance. Its unit of measure is the Ohm. And the Ohm gives us a very fundamental electrical theory and law. Ohm's law states that $E = IR$, with E being the units of volts, I units of amperes, and R units of ohms. Let us take a simple example. At our office work space, we have an electrical outlet of 110 V, 60 Hz power. Into this we plug an electrical device which has a resistance of 44 Ω. Rearranging algebraically to solve for I, we can calculate our current flow:

$$110 \text{ V}/44 \text{ } \Omega = 2.5 \text{ A}$$

How much power do we require? Using $P = EI$, we calculate P in watts = 110 V*2.5 A = 275 W. By knowing voltage, current flow, resistance, and some electrical theory, we can gain a better understanding of the electrical and electronic work being performed throughout the fastener manufacturing, distribution, and installation industries. We will wrap up this electrical theory induction with a few additional fundamental concepts. Another useful fundamental is Kirchhoff's law. The law provides a useful set of algebraic equations for the summation of the voltage drops around a circuit and the summation of current flows around the circuit and its branches. In learning how to calculate Kirchhoff's law equation, simple DC circuits of a battery and resistors are often used. We have discussed resistance in Ohm's law, circuits of resistors and batteries, and perhaps an on–off switch. These components can be assembled into circuits of series and parallel configurations in many possible combinations. Two 12-V batteries in series will provide 24 V of potential. These same two in parallel will provide 12 V but twice the current capacity in amperes. Resistors in series will be additive with the same current flow in ampere once the switch is closed. In parallel, the current will be divided between the two parallel branches in inverse proportion to their resistances. This explains the theory behind a short circuit. If a parallel path in an electrical circuit provides very low resistance, this path of least resistance will have very, perhaps, dangerously high current flow. Finally, our on–off switch can be assembled with two or more switches, either in series or parallel. Switches in series will only allow current flow when all switches are closed. Opening any one switch interrupts the circuit. This can be useful, for example, in an industrial operation where, owing to safety concerns, it is required to stop a circuit from operating (e.g., an electric motor) if one of several safety-related conditions are not being met. On–off switches in parallel can complete a path for circuit flow if any of them are closed. The classic example being the light switches at either end of a room. In the on position, the switch can be thought of numerically as a 1. In the off position, it is in the zero position. A circuit of switches opening and closing in a circuit of series and parallel paths represents the basics for understanding fundamental electronic programs. By providing states of ones and zeros (on and off), complex operations can be programmed with circuit conditions of "on–off" and circuits with "if–then" as well as "and–or" programming steps.

Other electrical components of interest are inductors and capacitors. Inductors are like the motor windings presented previously. The turns of an inductor provide a magnetic field. When these coils are arranged as in a transformer, the voltage can be stepped up or stepped down as needed. Examples are the 14,400 V transmission lines using line transformers first to compensate for voltage drops due to line resistance, which can increase and decrease with ambient temperatures. Transformers can also step the voltage down to 220 volts, 3 phase that enters our homes and places of business. It is at this point that we will wrap up our electrical technology for fastening review by introducing the important concept of the probable lead and lag of volts and amps in modern industrial circuits. A circuit with an alternating current flow as it flows through a resistor is in phase. The voltage and current rise and fall together. The fastener industry is loaded with many inductor-type loads. Here, the voltage and current become slightly out of phase. Capacitors, which are devices for storing electrical charge, are out of phase in the opposite direction. This leading and lagging of a circuit's phase condition, at any specific location in the circuit can be measured. The circuit for any piece of fastener machinery, or the entire factory for that matter, can be measured and expressed mathematically as its power factor. A power factor of unity, or 1, represents that the voltage and currents are in phase. This is an optimal electrical energy operating condition. Power factor of less than unity with current either leading or lagging voltage is expressed as a decimal percentage. Power factor gages, or meters as electrical gages are commonly called, can assist those with this technical information to use electrical energy most efficiently.

Reference

1. Selected notes on heat treating, developed during heat treating of fasteners.

8

Environmental Factors and Corrosion

8.1 Corrosion of Fasteners and Assemblies

Fastener corrosion represents a far greater loss to consumer, industry, and society as a whole than might first be imagined. Take, for example, a bridge crossing a river as part of an interstate highway system. If the various fasteners, bolts, washers, nuts, pins, and rivets that fasten the bridge corrode, society is faced with several costs. The bridge fastening systems might require a higher degree of maintenance than might otherwise have been the case if the corrosion rate for the specific fasteners were lower. Also, the corrosion might progress to the point of weakening the bridge, representing a very real and potentially catastrophic cost to all those who use or know someone who uses the bridge. Clearly, planning for the corrosive effects of the service environment and taking the necessary steps to provide the fastening systems with protection to control the rate of corrosion is an important factor in many applications. One look at a rusted-out automobile body is evidence enough of the potentially destructive forces that can shorten the useful life of fastener-joined engineered products. But unless anticipated, rust and other forms of corrosion or environmental weakening of fastener materials can cause less than planned for service.

Corrosion can be defined as the destruction or deterioration of a material caused by its reaction to its environment. Reaction is the operative term in understanding and controlling fastener corrosion. A good way to think about fastener corrosion is to model the fastener and its environment as a potential electrical circuit: some element in the fastener service environment combining with some element, perhaps iron or chromium, in the fastener material. There are two views anyone evaluating a fastener application can take with this corrosion model.

First, given a new application, good fastener engineering calls for a well-grounded knowledge of the intended service environment. Achieving this understanding, the fastener and fastened component materials should be specified based on their corrosion resistance to the effects of the service environment elements. Our concern is to set up the minimum corrosive reactions and corrosive loss of assembly/fastener materials.

Second, approaching an existing application, our concern first shifts to whether a material substitution in the direction of lower corrosion rates is possible. Next it is to create barriers to isolate the assembly materials from the service environment. Considering our model of the fastener corrosive application as an electrical circuit, in a new application we want to design a corrosive circuit with very low electrical potential and resultant current flow. With an existing, possibly corrosive, fastener application, our corrosion resistance goal shifts to building up resistance in the way of barriers to decrease to a trickle the potential current flow.

Fastener corrosion can be defined as the wearing away gradually as by the action of chemicals. A common form of fastener corrosion is rust. Other corrosion products might be chromium-chloride ions, oxides of aluminum, copper oxides, and many others. In some applications, the corrosion rate might be so slow that in the useful life of an assembly, no decrease in fastener size could be measured. In other applications, the corrosion rate might be rapid enough that catastrophic failure results from the clamped service overstressing the corrosion-reduced fastener area.

If an unprotected steel bolt is placed in service clamping two steel plates similar to the example provided in Chapter 2 but in a warm, moist environment, the iron in the steel would react with the oxygen in the water vapor to form iron oxide:

$$2Fe + 3H_2O = Fe_2O_3 + 3H_2$$

This equation of iron and water forming rust and hydrogen is a good starting point for an understanding of the corrosion mechanics of many steel-based fastened assemblies. Left unprotected and with no maintenance, steel will revert to rust in the presence of moisture. Atmospheric humidity, rain, and snow provide the water. Paths into the molecular iron of a steel part are present any time steel is exposed. Scratches, nicks, or worn and weathered fastener/assembly finish all present corrosion opportunities.

If the two parts of the right hand of the equation are kept separated, such as by a very robust finish, or a periodic coating of oil, iron oxide, commonly known as rust, is inhibited from forming. Fastener corrosion protection involves methods of keeping elements separated.

A model of a fastener corrosion system would look like the sketch in Figure 8.1.

Let us consider two fasteners, one from steel, zinc-plated, and the other 302 HQ stainless. The HQ, as an aside, designates the stainless as being of heading quality due to its having 3%–4% copper, which allows the stainless grains cold heading and toll threading tooling with better formability than standard type 302 stainless without copper.

Taking the two fasteners and installing them in an outdoor application, exposed to the elements, we would expect that over time the steel, zinc-plated fastener would begin to show first white corrosion products, then red corrosion

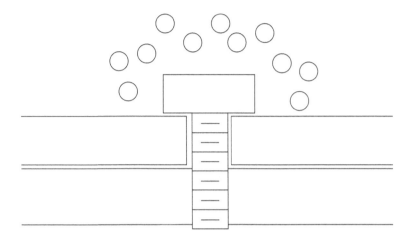

FIGURE 8.1
Model of fastener corrosion.

products, and finally red rust. Left unprotected, the iron in the steel will combine with oxygen in the air to form iron oxide, also known as rust. If we first coat the fastener with oil, electroplating, or some other protective coating that separates the iron from the oxygen, the process toward rust will slow.

Looking at the stainless fastener, if we have any exogenous (picked up in process) iron on the surface from manufacturing, it will also combine with oxygen in the air to rust. If our stainless fastener's surface is free from any exogenous iron, this eliminates rusting as a corrosion factor for the stainless fastener, but what about other types of corrosion? Here an understanding of 300 series stainless steel's corrosion character is needed. If we examined the surface of the stainless fastener from the time it was newly manufactured until it was in the installation for some time, we would see that with time this fastener material would form a darker and ever duller "skin," or oxide layer. What if we burnish the fastener in some abrasive media to polish it to a brilliant luster prior to installing it? The results would be the same, it would just take longer unless we separate the stainless metallurgical elements (iron, chromium, nickel, manganese, silicon) from the oxygen and other gases, in nature mostly inert nitrogen, but in marine or industrially hostile atmosphere perhaps sulfur and acidic gases that compose air.

This process of forming an oxide barrier is called passivation and can be formed naturally in air by 300 series austenitic stainless steel or can be brought about by treatment after manufacturing by immersion in an aqueous solution of nitric acid followed by a water rinse as called out in ASTM A380 and other cleaning and passivation specifications.

This barrier, whether naturally forming or applied as a supplemental coating or finish, is a key element of fastening corrosion resistance. Our stainless fastener example might serve us well in a service application environment at the seashore but be at risk of stress corrosion cracking in a pipeline application which included chlorine such as sour gas. Protection from the effects of corrosion should be a factor in the specification of fasteners and fastening systems. A checklist of corrosion protection steps is a useful tool for anyone contemplating assembly design to have at hand.

Fastener Corrosion Check List

1. Know the service environment.
2. Use corrosion-compatible materials.
3. Control the environment's effects.
4. Be wary of dissimilar material combinations.
5. Provide additional fastener area for possible corrosion loss.
6. Study fastener placement.
7. Do not provide crevices for corrosion.
8. Do provide for drainage of moisture.
9. Test assembled fasteners for corrosion.
10. Factor in corrosion effects on recycling.
11. Avoid excessive mechanical stresses and stress risers.
12. Design out uneven heat distributions.
13. Even out stress distributions.
14. Specify maintenance procedures.

Some discussion of these factors is in order. Key among these is, to the extent possible, to know the corrosive nature of the service environment. If the service environment can sometimes contain moist, salty air, or humid internal combustion engine-laden exhaust, the assembly can best be designed to "weather" the application if assembly design plans provide the best materials, configurations, procedures, and maintenance for these potentially corrosive elements.

In some environments, combining dissimilar materials cannot be avoided completely. However, judicious selection of the fastener and component materials by electromotive potential can help prevent a galvanic couple from occurring which could precipitate corrosion.

By knowing the most hostile service environment a fastened assembly will have to withstand and the fastener material's corrosion material loss rate in some units such as mils per year, additional fastener material can be provided to ensure against assembly weakening.

When establishing which sites in an assembly will be dedicated to the fastening function, regardless of the fastener type used, thought should be given to its placement in relationship to the service environment element. A good example in this regard are auto battery terminal fastenings. Once up on top of the battery, where corrosion and a loss of function was common, they now are often recessed into the side of the battery, partially protected from the corrosive battery vapors. It is also good fastener corrosion prevention to eliminate nesting areas for liquids and service dust and dirt. Nooks and crannies can provide fertile areas for corrosion to form and flourish. Proper drainage can often be provided by the placement of venting holes or paths for the free circulation of clean air. These access areas can also make cleaning and maintenance tasks easier.

Too often an assembly can be well wrung out from a clamp load efficiency viewpoint but not have the benefit of time-based field conditions testing. Service life in even the mildly corrosive environments found in many industrial and consumer applications can limit the product life and reusability of even finely engineered assemblies. While full-scale beta site testing may not be warranted in each application, laboratory and field testing that place corrosive conditions on a proposed assembly can yield valuable information on future customer use. Designed experiments and commercial laboratory equipment as well as selected field test applications can evaluate an assembly's corrosion resistance to heat, stress, and the chemical elements.

8.2 Corrosion Protection

Over the years, many materials and finishing processes have been developed to coat and protect fasteners from the damaging effects of environmental corrosion. One of the most popular and commonly used corrosion protection processes has been and continues to be electroplating. Plating is primarily an electrolytic process, meaning that electric current carries the cladding metal from a base electrode with one electrical charge to the fasteners which have been given the opposite charge. The plating material is "thrown" onto the fasteners by the current flow. On fasteners which have a wide range of surfaces and geometries, including head recesses and machine screw threads, this can present some real challenges. Meeting the opposing goals of adequate plating coverage without backing into a dimensional problem on some fastener feature due to overplating can be a challenge on some fasteners. Talk to a group of fastener people and plating horror stories are right up there among the difficulties in running their businesses.

It does not have to be that way on the fasteners that are in your assemblies provided some simple procedures are followed when manufacturing and plating the fasteners. First we need an understanding of the electroplating process. Many metals can be used to electroplate a substrate. Zinc is probably the most frequently used. It represents a good balance between cost and performance in a number of corrosion and fastener application factors. Cadmium has been used often and is called out on many military fastener specifications. Nickel, copper, tin, and brass are other electroplating materials.

The electroplating process can be either a rack or barrel operation. In rack plating, the products are hung on standoffs which carry them through the plating tanks. The contact point between part and standoff does not receive the "thrown" plating. Most fasteners are barrel-plated. The fasteners are carried through the plating tanks in perforated sided barrels which rotate so that all surfaces of the fasteners are contacted by the electrolyte in the tanks.

The plating line is composed of a series of tanks which first prep the surface of the fasteners and then apply the plating. It is a process with a trend of diminishing efficiency. By that I mean that the quality and quantity of plating coated onto the fasteners will decrease if the plating electrode is not replaced as it wears down and the electrolytic bath working fluid nit recharged when it weakens during plating. An improperly maintained electrolyte or electrodes result in less than optimum electroplating and less than desired corrosion resistance for the fasteners when installed in the application.

When you ask platers what they would like in the fasteners they receive to plate, it is that the surfaces be free from excess soot from heat treating or oils from manufacturing. They also ask for as much time as possible since theirs is usually the last processing step in the fastener-making process.

When they calculate the plating costs, they are not considering weight per unit as we consider in manufacturing. Rather they are calculating surface area since the amount of plating applied and hence plating operating costs are a function of fastener surface area.

Let us consider two cylindrical fasteners. A short, large-diameter one and a long, smaller-diameter part. They both might weigh the same amount per thousand parts, but the one with greater surface area will require more plating material for the same plating thickness.

Plating thickness is a function of the current density in the plating tank. As the current density is increased, more plating is thrown onto the fastener surfaces. Current density is a process control of the plating line operator and should be monitored in process.

The bath electrolyte can be either cyanide or acid depending on the plating material and electroplating process. The baths in operation can become contaminated if not cared for regularly with a decrease in the appearance and corrosion resistance of the fasteners.

Let us consider a simple clevis pin to be electroplated with zinc and given a supplemental immersion in a chromate tank. The plating coverage thickness and subsequent plating coverage would look like Figure 8.2.

Figure 8.3 shows the model for the fastener plating process.

This clevis pin example was a relatively simple geometry. If we consider a more sophisticated shape, such as a the square-socketed, shoulder screw with self-tapping screw threads for plastics found in the back of some computer keyboards, the variation of plating coverage becomes even more of a factor. Figure 8.4 shows a sketch of this fastener.

Clearly, in plating metal fasteners for corrosion resistance, we are dealing with two parameters which are at cross purposes unless they are properly reconciled. Corrosion resistance of plated metal fasteners is a function of plating thickness and porosity. Fastener function is based on geometric size. For

Thicker at outside surfaces

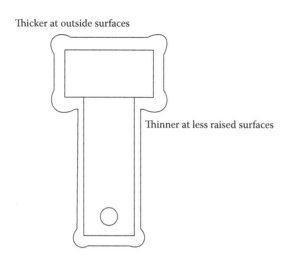

Thinner at less raised surfaces

FIGURE 8.2
Plating coverage of clevis pin.

example, if we increase the plating thickness on machine screw threads, they may not assemble in the mating threads. The plating may make the major, pitch, and minor diameters oversized. The plating grows the pitch diameter four times the plating thickness. Figure 8.5 shows a machine screw with a relatively standard 0.0002–0.0003 in. thick zinc plating with a supplemental yellow chromate finish. This is the type of threaded fastener one would find frequently in many types of electrical and electronic equipment. Several things are noteworthy. First is that the electroplating tends to build up more on the thread crests that form the thread major diameter. If we have a total buildup of 0.0003 in. per surface, our major diameter would increase $2 \times 0.0003 = 0.0006$ in.

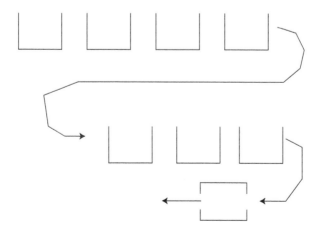

FIGURE 8.3
A typical plating process.

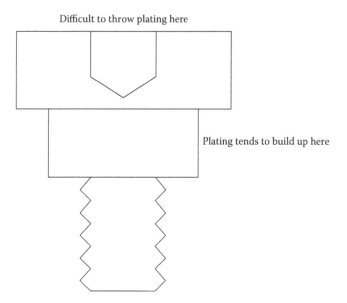

FIGURE 8.4
Plating coverage of a more complex geometry.

This is because we measure two major diameter surfaces with a micrometer's two anvils. Installed in the taped, mating internal thread, this extra 0.0006 in. requires clearance to avoid scuffing on the assembly and possibly losing some of the corrosion resistance. In the roots of the threads that form the minor diameter, a similar doubling of minor diameter occurs during plating.

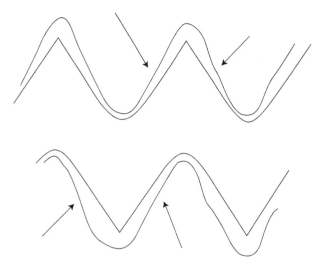

FIGURE 8.5
Plating build-up affecting thread pitch diameter.

On the pitch diameter, which is the working diameter of the threads, the pitch diameter, both single element and over many pitches, increases a multiple of four times, as a minimum, the plating thickness, on well-formed threads. Using a 0.0002–0.0003 in. plating thickness as an example, and estimating the size increase on the thread flanks, we will typically have closer to a 0.0002 in. thickness on these surfaces. The plated pitch diameter will be 4×0.0002 in. $= 0.0008$ in. larger in measured pitch diameter than when the threaded fasteners rolled off of the threading dies. Plating build-up should be considered in planning for assembly of tight fitting threads. Figure 8.5 shows this plating build-up phenomenon.

Some Other Often Used Fastener Finishes

	Advantages	Cautions
Electroless nickel	Superior corrosion resistance	Higher cost
Phosphate	Cost-effective	Oily
Dacromet	Superior corrosion resistance	Thickness higher cost
Black oxide	Good initial appearance	Low rust resistance
Zinc/black chromate	Good appearance	Hydrogen embrittlement
300/400 Series Stainless Steel		

The point here is that in providing electroplating or any commercial coating on threaded fasteners, plan on thread growth.

Figures 8.6 through 8.8 indicate some relative properties of some of the typical fastener grades of stainless steel.

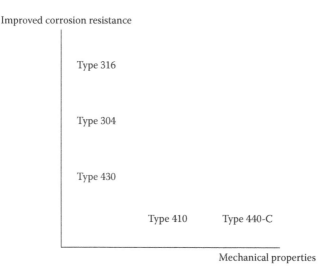

FIGURE 8.6
Chart of stainless fastener corrosion resistance versus strength.

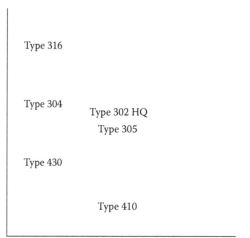

Increasing headability

FIGURE 8.7
Chart of corrosion resistance versus headability.

Noble (cathodic)
Platinum
Gold
Titanium
300 series stainless steel (passive)
400 series stainless steel (passive)
Nickel–copper alloys
Bronze
Copper
Brass
Lead
Stainless steel (active)
Cast iron
Steel
Cadmium
Aluminum
Zinc
Magnesium
Active (anodic)

FIGURE 8.8
Galvanic series in seawater at ambient temperature.

8.3 Corrosion Testing

A battery of test procedures are available to assembly engineers which can help avoid fastener corrosion problems. Corrosion testing can be performed at accredited testing laboratories, at the fastener manufacturing plants, and directly at assembly sites. It can be as simple as placing assemblies on a roof or at another outdoor site for long-term evaluation. It can also be as sophisticated as an environmentally controlled corrosion test chamber filled with corrosive fogs held under closely supervised temperatures and pressures.

One useful test for evaluating fastener corrosion rates and resistance is the salt spray test. These tests produce relatively quick indications of corrosion products that may form on the fastener and assembly surfaces exposed to a corrosive environment when installed in an assembly in the field. Neutral salt spray testing is used in fastener corrosion work in the same way the English or metric systems are used for measuring shank length. Standard ASTM salt spray test results are used to evaluate the corrosion resistance performance of materials and finishes on fasteners and fastened assemblies.

The neutral salt spray test is performed by atomizing a 5% by weight solution of sodium chloride and distilled water at a temperature held in a few degree range around 95°F in an environmentally controlled test chamber. A typical test chamber includes reservoirs for water and salt solution, a humidifying saturation tower, the salt fog-generating apparatus, a clean compressed air supply, and a controlled heating system.

The water used is distilled and checked frequently for the pH of the chamber condensate. Condensate collectors are such an important part of the test that their positioning and operation is a factor in the validity of comparative salt spray corrosion resistance data. The collectors should be away from the chamber walls and be at the same general level as the fastener test specimens. Figure 8.9 shows the plan of a laboratory-type salt spray tester.

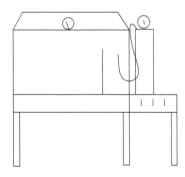

FIGURE 8.9
Salt spray tester.

The following lists the range of fastener corrosion tests available to the assembly designer:

Adhesion of metallic coatings on fasteners

Eddy-current measurement of plating thickness

Magnetic measurement of plating thickness

Pre- and postmicrometer measurement of specimens

Corrodent drop test

Microscopic measurement of plating thickness

Strip and weigh method of plating thickness

Salt spray testing

Humidity testing

Stress corrosion

Stainless steel intergranular corrosion nitric acid

Stainless steel intergranular oxalic acid etch

Hydrogen embrittlement testing

It is worth adding a note at this point on fastener finish and corrosion testing. The goal in all such testing should be to provide information on the effectiveness of the plating or finish to protect the fastener in the installation environment. The closer the tests come to providing accurate information in a safe, timely, and economical manner, the better. While the above list represents some common fastener industry tests, individually developed tests for a specific fastening application can sometimes provide a focused view of fastener finish performance improvement areas.

8.4 Some Thoughts on Sustainability

In the second edition of this book, I want to introduce the importance and impact of sustainability to those involved with the manufacture and use of the wide range of fasteners used throughout the industrialized world. Too often in the past, manufacturing and assembly practices were often employed in these areas without much thought given to environmental effects, especially on future generations using these same natural resources. We have seen a much greater awareness of the importance of recycling fastened assemblies, especially in traditional industrial industries such as automobiles and house appliances where turnover is often in 5-, 10-, and 20-year product life cycles. When the first edition was written, large "car crushers" were capable of grinding up a scrapped automobile whose drive train as well as other re-buildable components were removed. This ground scrap metal could then be remelted and fed to be alloyed into future new steel raw materials. Similarly, the injection-molded plastic components found in autos and appliances could be reground as feed stock for new injection-molded parts. Electronics, especially mobile devices, has enjoyed widespread consumer adoption. Their sustainability in terms of recycling materials is less certain. Unlike more clearly recoverable materials such as steel, rubber, glass, and plastic, modern devices contain in addition to those materials elements such as rare earth metals. These items have entered a rapid growth phase in production. It remains to be seen how their various components are recovered and reused once their productive lives as useful products are over. I have observed that like automobiles, these mobile electronic devices move from production facilities, through sales channels to first the new device user then to a second user. Also as in the automobile life cycle, it will be the economic incentive of the recoverable materials that will provide at least part of the not just bury these devices in a land fill where pollution the area's soil is a strong probability. Government motivated consumers can also play a pivotal role in encouraging responsible reuse and emphasis of sustainability of both the assembly materials and the environments through which they reside throughout their assembly life cycle.

8.5 Application Engineering Project Management

For a period stretching over quite a time period, I presented a workshop called Fastener Fundamentals. During these workshops, the training goal was to impart much of the foundation technology and techniques used to work with fastener manufacturers and consumers. The thought being that thorough knowledge and efficient practice would lead to fastened assemblies that were equally profitable for consumers and producers alike. The preceding seven chapters have presented brief reviews of the application engineering core subjects that are important to know. I encourage the reader to explore at their own pace each of these subjects in more depth. Think of it as exercise for the technical mind. Many college and university courses are now available online. I found it very useful to explore these topics in concert with whatever fastener application problem I was working on at that particular time. Following is an outline of some of the Fastener Fundamentals Workshop materials:

Designing and Assembly with Fasteners

Assembly Component Fundamentals
Useful definition
The application engineer's role
How to internalize technical information

Torque/friction/preload

Elongation

Fatigue strength

Impact strength

Technical Sales Technique

Understanding the customer

Setting mutual goals

Doing your homework

The scientific method

Numerical approaches

Getting support

The proper mindset

As you work through your development, or as a refresher, making your own personal notes of the above list will increase the depth of your fastener and applications understanding.

8.6 Thermodynamics

As an electrical apprentice, the classwork and theory I received started my technical interest in the electrical and electronic technology reviewed at the end of Chapter 7. On the ships in the Navy Yard, I was fortunate to be able to witness another fundamental technology, in action. That area of technology is thermodynamics. We can think of thermo as the study of heat. What is heat. Heat is a form of energy which can be expressed in scientific units such as calories and British Thermal Units (BTUs). A calorie is something we hear a lot about in terms of nutrition. This is a good starting point in that our bodies convert food into heat which our bodies need and use. In engineering terms, a calorie is the heat energy required to increase the temperature of one gram of water one degree centigrade. Let us consider how and understanding of thermodynamics meshes with the application of fastener applications. I can think of two ways. First all fasteners are in environments where there is a presence of heat energy, be it the higher heat of a piece of fastened equipment operating at temperature or a structural fastened joint existing in a remote artic site with an absence of heat to any significant degree. The second is thermodynamics emphasis on a systems view of a system and its process. This is fundamental to the practice of fastener applications. The term "thermodynamics" is derived from "thermo," which means heat, and dynamics, which means "motion." It means heat in motion. And the fastener applications we will be fortunate to work with will all have some degree of thermal change. Viewing this as a change over time and mentally making a system of the complete fastened assembly and its operating environment provides us with a powerful set of tools. Two of the key laws that govern thermodynamics are the first and second laws of thermodynamics. The first establishes the fact of the conservation of energy. Heat is stored energy. In fastener manufacturing, the term "cold heading," which we described, is in fact a process which increases the heat content of the cold-headed blanks. You cannot catch a blank coming out of the discharge chute of a cold header producing screw blanks without some protection for your hands do to the heat increase. A thermodynamic term I read in my notes on thermodynamics while preparing this section is the term adiabatic. It is a rare term as very few processes, certainly in the fastener world, are adiabatic, which means no heat transfer. Staying with the first law of thermodynamics, the conservation of energy provides a tool for analyzing systems with respect to energy. We are all becoming more energy-conscious due to both economic and environmental reasons. Classic thermodynamics was an early user of these "environment impact" studies. This technology, which was high tech and the area of a select few intellectuals and scientists around the time the United States was gaining independence, provided the knowledge for steam engines, internal combustion engines, jet turbines, and finally out space rocket propulsion systems. It starts with a definition of a system. Systems can be open or closed. An example of an open system is a modern jet engine. Large quantities, or more technically correct, a specific mass of air moves in, through, and out of this thermodynamic system as heat is transferred from burning jet fuel to hot jet exhaust providing the thrust. A closed system which confounded engineers for many decades before it was solved to increase people's quality of life, and use a lot of fasteners, is the refrigeration cycle. The working fluid in the refrigeration cycle, as found in an air conditioner or refrigerator, is a special refrigeration that operates in a closed system. It is compressed and expanded in a thermodynamic cycle which allows it to absorb heat from a higher content reservoir and transfer it to a reservoir with a lower heat content. This is a classic example of thermodynamic theory at work. The measurement of heat introduces some new measurement scales. We are familiar with the Centigrade and Fahrenheit temperature scales. Thermodynamics introduces the Rankin and Kelvin scales, which are used in thermodynamics calculation as well as other scientific work. The second law makes use of these scales. The second law of thermodynamics introduces the student and reminds the practitioner that systems are composed of a total energy content which can be divided into usable energy and that which is un-usable. The second law states that systems tend to the condition of maximum unusable energy. The unusable energy is given the term "entropy" and the symbol S.

Total energy is U and usable energy is H. Systems over time will maximize S unless work is performed on the system: $U = H–S$. This is a power concept and exceedingly useful for the fastener specialist. Iron-based fasteners rust unless the iron is separated from the oxygen in the air when newly installed, and periodic protective work is performed with time. It also reinforces a fact commonly come across in engineering, which is system losses. In any process, or system cycle, energy is transferred, work is performed, and a percentage of the total energy is unusable or waste with respect to the work being performed. The ideal gas laws setting the physical property relationship of temperature, pressure, and volume of a working fluid such as a gas in the refrigeration unit helps us understand that as pressure of the working gas increases, volume decreases and temperature rises. Reversing this cycle, by expanding the volume, results in a decrease in the heat, as measured by a temperature scale and a drop in pressure. Developing a knowledge base of thermodynamic laws, theory, principles, and operating examples will expand a technical person's systems mind view. This ability, when developed, translates well to the study, development work, and ability to improve many other classes of systems, such as fastening systems, processes, and assemblies. It all started for me when observing steam turbine engines, which are held together with a lot of special fasteners, driving electric generators on navy ships. I then underwent a more formal and numerically rigorous training by preparing and sitting for the professional engineering examinations. This is a good juncture to discuss the importance of being a well-rounded practitioner and a student for life. It is easy, in a busy career, to set aside the search for continual learning and of expanding oneself in new areas of interest. In early and mid-career, I was able to squeeze in time for technical refresher and review classes as well as to conduct an independent study on certified manufacturing engineering, fundamentals of engineering, and professional engineering examinations. These activities guide you to learn and hone new skills and technologies. In this later point in life, music and learning to play an instrument have captivated me. Being a "civilized" engineer by being well read and well rounded helps your professional life in solving fastener application problems. The time tested adage "learn something new every day" continues to ring true. In these first eight chapters, we discussed statics, dynamics, strength of materials, chemistry, engineering economics, electrical theory, and thermodynamics. It is hoped that these brief sections will interest and encourage you to explore further as application and time allow. As a final concept on heat, thermodynamics involves the study of heat transfer. There are three forms of heat transfer: conduction, convection, and radiation. Conduction is the transfer of heat through a material. In a fastener application, a steam turbine bolt will conduct heat from the steam through the turbine housing and along its screw threads, shank, and head. If not planned for in the design and application of these fasteners, repeated heat transfer can make removal of the fastener very difficult. With conduction, air passing over the heads of bolt heads on a cooling tower will act as convection currents to transfer the heat conducted up through the bolt. The air acts as a working fluid to transfer heat from higher to cooler areas. Finally, the metal fasteners in a residential heating installation will radiate heat from their surfaces to the adjoining installation environment. Conduction, convection, and radiation are important heat-transfer concepts. The rate of heat transfer is a function of the temperature gradient in the system with larger "delta Ts" driving larger and differing rates of heat transfer as measured in heat units such as calories or BTUs, with 252 calories equaling 1 BTU.

As a point of interest on fluids at work in a system, the steam turbine I witnessed as an apprentice was driven by superheated steam. Heat added to water raises its temperature. Once at the boiling point, water experiences a phase change to steam. If that steam is contained in a pressure vessel and heat is continued to be added, the pressure increases and the "superheat" becomes a powerful force acting on the steam turbine blades as they rotate at high speeds. The rotating turbine shaft can turn a generator set of windings through a magnetic field, inducing strong electrical current in the rotor winding which can be collected by the commutator brushes and sent on through a switch board and on to do useful work. This is thermodynamics at work.

Suggested Reading

1. *Carpenter Technology, Heading Hints*, Carpenter Technology, Reading, PA, Copyright, 1985.
2. *Carpenter Technology, Corrosion Causes and Control*, Reading, PA, Copyright McGraw-Hill, New York.

9

Assembly Analysis

We have established the fastening fundamentals of strength, appearance, and reusability. In Chapter 1, we touched the importance of controlling the number of fasteners used in an assembly and the value of combining assembly functions in fewer fasteners for reduced assembly cost. In the 18–82 principle, 82% of assembly costs are installation and transport functions, with only 18% representing the fastening purchasing cost. This approximately 4.5:1 ratio of fastening costs is important to weigh against the design integrity requirements. Specifying an efficient fastening system requires both qualitative and quantitative judgments on competing products and solutions. By staying current with the fastening industry and employing some of the quantitive analysis we will present, the fastener specifier should be better equipped to make assembly specification decisions.

9.1 Circular Areas

9.1.1 Pins, Shoulders, Shear Areas

Circular fastener areas can be determined by using the equation:

$$A_c = \Pi D^2 / 4$$
$$= 0.7854 \, D^2$$

As an example, let us select a pin diameter of 0.125 in. and find the shear area.

$$A_c = 0.7854 \, (0.125 \text{ in.})^2$$
$$A_c = 0.7854 \, (0.0156 \text{ in.}^2)$$
$$A = 0.0123 \text{ in.}^2$$

9.1.2 Hexagonal Area

If the area is a hexagon, we use the equation:

$$A_h = 3.4641 \, (r^2)$$

where r is half the across flats dimension.

Let us calculate a hex head area with 0.125 in. across the flats. The value of r would be 0.0625 in.

$$A_h = 3.4641 \ (0.0625)^2 \ \text{in.}$$
$$A_h = 0.1135 \ \text{in.}$$

For any fastener geometry, which in many applications would be either circular or hex shaped, we find the area by using the area equations and solving for the specific dimensions of the fastener.

9.1.3 Circular Volume

To find the volume of a fastener, we would multiply the area by the height of the fastener or fastener section. The equation is

$$V_c = 0.7854 \ (D)^2 h$$

For example, if our 0.125 in. circular diameter has a height of 0.093 in., we would solve for volume as follows:

$$V_c = 0.7854 \ (0.125)^2 \ \text{in.} \ (0.093) \ \text{in.}$$
$$V_c = 0.7854 \ (0.0156 \ \text{in.}^2) \ 0.093 \ \text{in.}$$
$$V_c = 0.7854 \ (0.0015 \ \text{in.}^3)$$
$$V_c = 0.0011 \ \text{in.}^3$$

9.1.4 Hexagonal Volume

Using a similar technique, we would multiply by the hex head height to obtain the hex head volume. Our equation would be

$$V_h = 3.4641 \ r^2 h$$

Using our previous example and a head height of 0.093 in., we would solve as

$$V_h = 3.461 \ (0.0625)^2 \ \text{in.} \ (0.093 \ \text{in.})$$
$$V_h = 3.461 \ (0.0039 \ \text{in.}^2) \ 0.093 \ \text{in.}$$
$$V_h = 3.641 \ (0.0004 \ \text{in.}^3)$$
$$V_h = 0.0013 \ \text{in.}^3$$

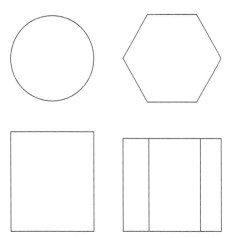

FIGURE 9.1
Circular and hexagonal areas and volumes.

It is interesting to note that with these two examples, the hex has approximately 10% more volume due to the six hex corners. The diametral material such as head diameters are a squared function and increase volume more quickly than when material is added to the height (see Figure 9.1).

9.1.5 Volume of a More Complex Shape

9.1.5.1 Fastener Geometry

If we have a fastener with several geometries, say a hex head that also has a phillips recess, an unthreaded shoulder, and a smaller threaded section, a composite volume could be estimated. A typical equation would be

$$V_T = V_{head} - V_{recess} + V_{shldr} + V_{thread}$$

Let us set hex head across flats = 0.125 and head height = 0.100 in.

$$\begin{aligned} V_{head} &= 3.464\, r^2 h \\ &= 3.464\,(0.0625 \text{ in.})^2\, 0.100 \text{ in.} \\ &= 3.464\,(0.0039 \text{ in.}^2)\, 0.100 \text{ in.} \\ &= 3.464\,(0.0004 \text{ in.}^3) \\ V_{head} &= 0.0014 \text{ in.}^3 \end{aligned}$$

As an approximation of cross-recess displaced volume, 15% is a good estimate.

$$\begin{aligned} V_{recess} &= 0.15\,(0.0014 \text{ in.}^3) \\ V_{recess} &= 0.0002 \text{ in.}^3 \end{aligned}$$

Let us use 0.100 in. for our shoulder diameter and 0.90 in. for shoulder length:

$$V_{shldr} = 0.785\, d^2 h$$
$$= 0.7854\, (0.100)^2 \text{ in. } (0.090) \text{ in.}$$
$$= 0.7854\, (0.010 \text{ in.}^2)\, 0.090 \text{ in.}$$
$$V_{shldr} = 0.0007 \text{ in.}^3$$

It is interesting to note as we proceed with this complex geometry volume solution that volume grows faster with increases in diameter as opposed to increases in length due to the squaring exponent of the diameter.

For our final section, we calculate the volume in the threaded area. Here we use the roll thread diameter. For our example, we will use a machine screw thread of size #4-40, a number 4 diameter with 40 threads per inch. Tables 9.1 and 9.2 show the roll thread diameters for the more frequently specified inch and metric system fasteners. If a less common, special thread geometry or spaced, self-tapping thread is used, you can select a diameter as a first approximation for this calculation—a diameter equivalent to the theoretical

TABLE 9.1

Inch Roll Thread Diameters

Size	Blank Diameter	Size	Blank Diameter
#0-80	0.050		
#1-64	0.061	#1-72	0.062
#2-56	0.072	#2-64	0.074
#3-48	0.082	#3-56	0.085
#4-40	0.092	#4-48	0.096
#5-40	0.105	#5-44	0.108
#6-32	0.116	#6-40	0.120
#8-32	0.140	#8-36	0.144
#10-24	0.160	#10-32	0.167
1/4-20	0.214	1/4-28	0.224
5/16-18	0.274	5/16-24	0.283
3/8-16	0.332	3/8-24	0.345
7/16-14	0.288	7/16-20	0.403
1/2-13	0.447	1/2-20	0.466
9/16-12	0.506	9/16-18	0.525
5/8-11	0.564	5/8-18	0.587
3/4-10	0.683	3/4-16	0.707
7/8-9	0.800	7/8-14	0.827
1-8	0.916	1-12	0.944

TABLE 9.2

Metric Roll Thread Diameters

	Blank Diameter	
Size	inches	mm
M1.4	0.044	1.12
M1.6	0.0525	1.33
M1.7	0.0540	1.37
M1.8	0.0603	1.53
M2 × 4	0.0669	1.70
M2.2	0.0735	1.87
M2.5	0.0853	2.17
M2.9	0.0990	2.51
M3 × .5	0.1036	2.63
M3.5	0.1207	3.06
M4	0.1378	3.50
M4.5 × .75	0.1562	3.97
M5	0.1746	4.43
M6	0.2088	5.30
M7	0.2480	6.29
M8 × 1.25	0.2812	7.14
M8 × 1	0.2873	7.29
M10	0.3534	8.97
M10 × 1.25	0.3598	9.13
M12	0.4217	10.7
M14	0.4381	11.12
M14 × 1.5	0.5105	12.96
M16	0.5765	14.63

pitch diameter. This is the diameter approximately halfway between the thread crests and roots:

$$V_{thread} = 0.7854 \,(0.092)^2 \text{ in. } (0.312 \text{ in.})$$
$$= 0.7854 \,(0.0085 \text{ in.}^2) \, 0.312 \text{ in.}$$
$$= 0.7854 \,(0.0026 \text{ in.}^3)$$
$$V_{thread} = 0.0021 \text{ in.}^3$$

Summing up the individual section volumes for our cross-recess, hex head shoulder machine screw, we solve for the total volume estimate for the part:

$$V_T = V_{head} - V_{recess} + V_{shldr} + V_{thread}$$
$$V_T = 0.0014 \text{ in.}^3 + 0.0002 \text{ in.}^3 + 0.0007 \text{ in.}^3 + 0.0021 \text{ in.}^3$$
$$V_T = 0.0040 \text{ in.}^3$$

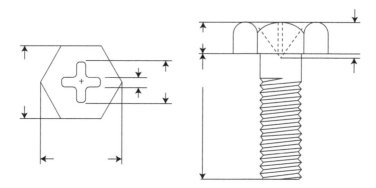

FIGURE 9.2
Volume of a more complex geometry.

Any fastener can be sectioned into geometric sections and solved for the volume that fastener contains. It is important to keep in mind that the outside and inside corners are calculated as volumes for full sharp corners where in fact fastener corners will contain some radii which decrease actual volume slightly. Also, with an accurate scale, part(s) can be weighed for average part weight and divided by density to gain part volume. Figure 9.2 shows the part.

The roll thread diameter for any given combination of thread including angle, root profile, helix angle, number of thread starts in the case of multiple lead threads, and minor/major diameters, along with the tolerance classes, can be found in the industrial, federal, or international thread standards. In manufacturing threads in the fastener factory, a set of blank diameters is developed and can vary slightly for the same thread size by material. For example, rolling the same thread in 300 series stainless steel may require a blank a few tenths different in diameter than if rolled from low-carbon steel.

Figure 9.3 shows an illustration of the roll thread diameter/threaded fastener relationship.

9.1.6 Calculate Part Weight, Individual, Per Hundred, Per Thousand

Let us take the preceding example volume of a more complex fastener geometry. We want to know what the weight of our part will be and how much raw material we will need. We will select low-carbon steel, type 1018 as our fastener material:

$$1018 \text{ Steel Density} = 0.28 \text{ lb/in.}^3$$

$$V_{TOTAL} = 0.004 \text{ in.}^3$$

To find the part weight, we multiply part volume times fastener material density:

$$W(\text{lb}) = V_{TOTAL} \text{ (in.}^3) \times \text{Density (lb/in.}^3)$$

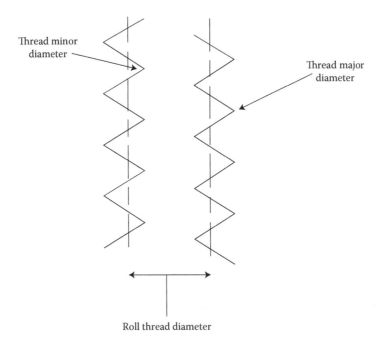

FIGURE 9.3
Roll thread diameter of a threaded blank.

Using our example data

$$W = 0.004 \text{ in.}^3 \ (0.28 \text{ lb/in.}^3)$$
$$W = 0.0011 \text{ lb}$$

our part weighs 0.0011 pounds each. To find the weight of a hundred, a thousand, or a hundred thousand, we would use the appropriate multiplying factor:

$$1C \ part = 0.11 \text{ lb}, \quad 1K \ part = 1.1 \text{ lb}, \quad 100K = 110 \text{ lb}$$

In actual practice, we would add 1–5% raw material for machine setup and hopefully small amounts of scrap. We could use the unadjusted weight with some variance factor for processes such as heat treating and shipping and handling computations (see Figure 9.4).

Fastener volume × material density = part weight

FIGURE 9.4
Equation form for fastener weight.

9.1.7 Calculate Fastener Elastic Elongation

It is fundamental to the understanding of threaded fastener operation to know that under tensile loading, threaded fasteners elongate. If we want to approximate the amount of stretch that a given threaded fastener has for a given tensile load, we can use the elongation equation

$$e = P \times L/A \times E$$

where P is the tensile load, L is the grip length, A is the stressed area, and E is the fastener material's modulus of elasticity. Let us use as an example a grade 5 hex head cap screw, 5/16 × 3 in., preloaded to a tensile load of 6100 pounds. We will state that it has a gripped length in the assembly of 1.5 in. What would be the approximate increase in length in the fastener's grip length? The fastener is threaded to the head:

$$E = 6.100 \text{ lb } (1.5 \text{ in.})/ \ 0.0616 \text{ in.}^2 \ (30 \times 10^6 \text{lb/in.}^2)$$
$$E = 9150 \text{ lb in.}/1,848,000 \text{ lb}$$
$$E = 0.005 \text{ in. or change in length (Delta L)}$$

Our 1.5 in. grip length will elongate approximately 0.005 in. under load to 1.505 in. This elongation under load is an important concept. It is the spring-like nature of threaded tension fasteners to stretch elastically that enables them to clamp.

It is also worth noting that the elongation for a given preload, such as our 6100 pounds, is a direct function of grip length. As grip length increases, so does elongation. For short grip lengths, less elongation occurs, which can result in substantial preload percentage loss if even slight joint relaxation occurs. Figure 9.5 shows the problem sketch.

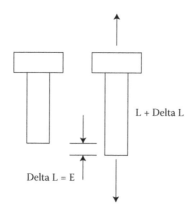

FIGURE 9.5
Fastener elongation.

9.1.8 Calculate Upset Volume

A limit to the head diameter and head height of cold headed fasteners exists as a function of the number of "blows" or times it is worked on by the cold heading punches. A typical screw or rivet cold header would use two punches and a single die. For these calculations, we define "one diameter" as a length of wire equal to its diameter. For example, one diameter of 0.092 in. wire would be 0.092 in. long. Each blow can upset approximately 2 1/4 diameters of material. A two blow header can upset roughly 4 1/2 diameters (equivalent lengths) of a given diameter wire. As an example, let us calculate the following:

A #4-40 threaded fastener with a round cylindrical head, 0.200 in. diameter, and 0.085 in. head height is to be cold headed. How many diameters must be upset from the 0.092 in. wire?

$$\text{Volume} = 0.7854 \, (0.200)^2 \text{ in. } (0.85) \text{ in.}$$
$$= 0.7854 \, (0.040 \text{ in.}^2) \, 0.085 \text{ in.}$$
$$= 0.7854 \, (0.0034 \text{ in.}^3)$$
$$= 0.0027 \text{ in.}^3$$

$$\text{Length of wire needed} = \text{volume/area}$$
$$= 0.0027 \text{ in.}^3 / 0.7854 \, (0.092)^2 \text{ in.}$$
$$= 0.0027 \text{ in.}^3 / 0.7854 \, (0.0085 \text{ in.}^2)$$
$$= 0.0027 \text{ in.}^3 / 0.0066 \text{ in.}^2$$
$$= 0.4017 \text{ in.}$$

Diameters to cold head this geometry from this wire size:

$$\text{Upset diameters} = \text{Wire length/wire diameter}$$
$$= 0.4017 \text{ in.}/0.092 \text{ in.}$$
$$= 4.4 \text{ diameters}$$

This could be cold headed on a two blow, single die cold header. Figure 9.6 shows the problem drawing.

9.1.9 Find the Largest Head Diameter per Upset

Suppose we want to find the largest cold-headed fastener diameter we can upset with 4 1/2 diameters in the head from a given wire size. Let us take a #8-32, using 0.138 in. wire. We will say the head height is 0.90 in. We would first calculate the volume in a 4.5 diameter upset, then solve for head diameter as follows:

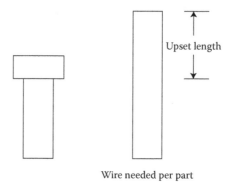

FIGURE 9.6
Normal upset fastener cut-off.

4.5 diameter upset volume = 0.7854 (0.138 in.)^2 4.5(0.138 in)

= 0.7854 (0.019 in.^{2}) 4.5(0.138 in.)

= 0.7854 (0.019 in.^{2}) 0.621 in.

= 0.7854 (0.0118 in.^{3})

= 0.0093 in.^{3}

Then we set the volume equation equal to this volume and solve for the diameter:

0.0093 in.^{3} = 0.7854 D^2(0.090 in.)

0.0093 in.^{3}/0.7854(0.090 in.) = D^2

0.0093 in.^{3}/0.707 in. = D^2

0.1314 in.^{2} = D^2

Then we take the square root of each side of the equation to find the largest possible head diameter:

$$\sqrt{0.1314} \text{ in.}^2 = \sqrt{D^2}$$

0.3625 in. = D

The largest head diameter we could approximately form using a 2-punch, single-die conventional cold head and most typical cold-heading fastener materials with a 0.90 in. head height would be 0.363 in. With a recess and radii on the head O.D., this could be increased slightly. The problem is illustrated in Figure 9.7.

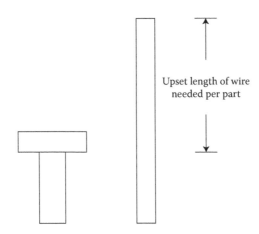

FIGURE 9.7
Oversize upset fastener cut-off.

9.1.10 Calculating a Compound Fastener Shape

Step #1	Separate the fastener into simple geometric shapes
Step #2	Calculate each volume
Step #3	Add the solid volumes and subtract any hollow volumes such as drives, recesses, holes, etc.
Step #4	Test the calculated volume by comparing it to known fastener volumes and also dividing actual part weights by their materials density

Example:

Volume 1	Volume of head, solid hexagon
−Volume 2	Volume of screwdriver slot, hollow rectangle
+Volume 3	Volume of shoulder, solid circular cylinder
+Volume 4	Volume of threaded section, solid circular volume, use roll thread diameter

Total = Compound Fastener Volume

9.1.11 Estimating Screw Slot Torque

Fastener data: Slotted machine screw

Slot depth = 0.037 in., solt width = 0.045
Head diameter = 0.260 in. #6-32 Thread
Fastener material yield strength = 45,000 psi

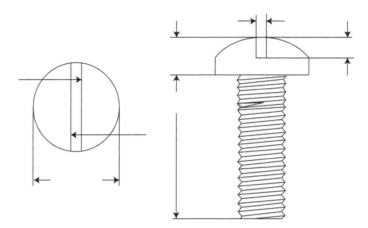

FIGURE 9.8
Screw slot torque transmitting ability.

Screw driver wall area approximates 30% of total wall area, as shown in Figure 9.8:

$$\text{Wall area} = 0.3(0.260 \text{ in.}) (0.037 \text{ in.})$$
$$= 0.0029 \text{ in.}^2$$

Drive force which can be resisted prior to onset of wall yield:

$$45,000 \text{ psi } (0.0029 \text{ in.}^2) = 130.5 \text{ lb}$$

Estimate of limiting torque of screw driver slot:

$$\text{Torque} = 0.130 \text{ in. } (130.5 \text{ lb})$$
$$= \text{Approximately 17 in. lb}$$

This would apply an approximate clamp load of 600 pounds of force and develop stresses in the area of 58,000–60,000 psi in the screw material. From experience, these appear to indicate a slot of sufficient geometry to handle the torque it is expected to see (see Figure 9.8).

9.1.12 Calculating Strip-to-Drive Ratios

When a self-threading screw thread is used in an application, the torque–tension relationship is altered. Torque is needed in larger amounts during the starting and running-up revolutions of the self-threading screw since additional work is being performed in displacing the material on the walls of the site hole to generate the mating internal threads. Since this usually involves relatively larger areas of the self-threading geometry rubbing against the site material during this thread forming, the torques usually ramp up during the starting revolutions and then stabilize or drop off slightly until clamp up occurs.

The limiting strength of the newly threaded joint can be either the screw breaking or, more likely, the internal threads stripping. By making a ratio of the strip-to-drive torques, a measure of the margin between these torques can be an indicator of how user-friendly the self-threading design is in the specific site.

Let us consider two candidate thread formers:

	Self-Threader Design A	Self-Threader Design B
Drive torque	20 in. lb	25 in. lb
Strip torque	40 in. lb	80 in. lb
Strip/drive ratio	2:1	>3:1

Although design A requires less drive torque to form threads, the distance in torque between the thread forming and joint failure is much lower than in design B. An overrunning of the tightening tools or other installation variables could result in damaged or scrap assemblies. It is desirable to evaluate not just the drive torque but the strip-to-drive ratio of self-threading screws in any specific site to determine the torque allowance that is highlighted by determining the strip-to-drive ratios. Figure 9.9 illustrates the strip/drive function.

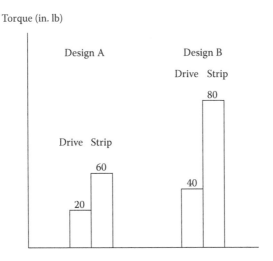

FIGURE 9.9
Strip-to-drive rations of self-threading screws.

9.1.13 Evaluating Drive Torque/Horsepower

Many people in the fastener industry for any period hear the question "What torque do we tighten these to?" The fact is that fastener tightening torque is really a misused and misunderstood concept for many fastener industry members. Torque is simply a force that causes rotation. Let us say we have a 20-pound suitcase on the floor constructed with two handles, one on top and one on the side, and further that the top handle is in line with the suitcases' center of gravity. A pulling force larger than 20 pounds will lift the case off the floor, but a 20-pound force on the side handle will first cause the suitcase to rotate. This rotation is the result of torque. With threaded fasteners, we need this rotation to thread the fasteners and mating assembly parts together. We may want to know how much horsepower, or fraction thereof, we need for a given fastener torque (Figure 9.10).

Our equation would be as follows:

Required torquing power = (0.0019) drive torque/s

As an example, calculate the fractional horsepower for a fastener drive using 30 ft-lb/s.

$$\text{Power required} = (0.0019)\ 30\ \text{ft-lb/s}$$
$$= 0.055\ \text{hp}$$

This is a little over 1/20th of a horsepower, or 40.7 W. With a 110 V AC motor, the current would be approximately

$$40.7\ \text{W} = 110\ \text{V}\ (\text{I})$$
$$\text{I} = 40.7/110$$
$$\text{I} = 0.4\ \text{A}$$

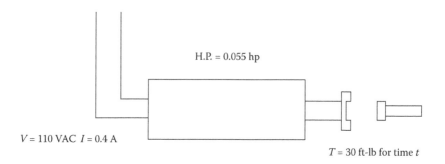

H.P. = 0.055 hp

V = 110 VAC I = 0.4 A

T = 30 ft-lb for time t

FIGURE 9.10
Fastener driving horsepower.

9.1.14 Calculating a Knurl Pitch

Often a knurl can be rolled onto the shank or head of a fastener to either make handling easier or provide a surface for pushout and torque resistance when press fit into a cored, drilled, or molded hole. There are many knurl designs that can be manufactured. Diamond, herringbone, and straight knurls are among the most common. It is often desirable to be able to determine what pitch or teeth per inch can be rolled onto a round cylindrical fastener. For any given circular diameter, there are a number of teeth that will index evenly onto the part. Using a pitch less than optimal will result in slivering when rolling. The equation is

$$Nt = 3.1416(Bd)/P$$

$$\text{where } Nt = \text{Number of teeth}$$
$$Bd = \text{Blank diameter}$$
$$P = \text{Pitch}$$

As an example, a round cylindrical blank of 0.318 diameter calls for a straight knurl with 12 teeth around the circumference. What pitch/TPI knurl should be used?

$$12 = 3.1416(0.318)/P$$
$$P = 3.1416(0.318)/12$$
$$P = 0.0833$$

This is the spacing from crest to crest of the knurl teeth. To find threads per inch, TPI, we divide this into 1 in. as follows:

$$TPI = 1/0.0833$$
$$= 12$$

Figure 9.11 shows the knurling diameter and pitch.

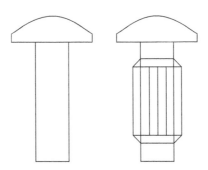

FIGURE 9.11
Calculating knurl dimensions.

9.1.15 Finding a Screw Diameter from a Target Clamp Load

If a clamp load is determined and you want to find a diameter that will apply the target load and remain below its yield strength, the following calculations can provide a useful model:

Target clamp load 125 lb

Screw material yield strength 50,000 psi
Area of highest stress for the application is in the screw threaded area. Calculate thread stress:

$$\text{Select a design factor} = 2$$
$$A/2 = 125 \text{ lb}/50,000 \text{ psi}$$
$$A = 0.005 \text{ in.}^2$$

Calculate a screw thread stressed diameter

$$A = 0.005 \text{ in.}^2 = 0.7854 \, d^2$$
$$0.005 \text{ in.}^2/0/7854 = d^2$$
$$0.0064 \text{ in.}^2 = d$$

$0.080 \text{ in.} = d$, or a diameter approximately equivalent to a #2 or M2 diameter machine screw. Figure 9.12 shows the problem sketch.

9.1.16 Calculating Pin Shear Strengths

Often when the service load to be carried is one of shear, a pin-type fastener or fastener feature is desirable to carry these loads. Shear strength

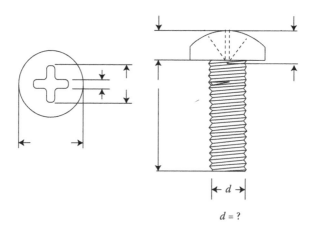

$$d = ?$$

FIGURE 9.12
Target clamp load screw diameter.

measurements, carried out on a tensile tester with suitable fixtures, and fastener shear strength calculations can be for either single or double shear strength. Let us look at a solid pin of 300 series stainless steel with a shear strength of material of 48,000 psi. We will assign it a diameter of 3/16 in. Our shear strengths would be calculated as

$$\text{Shear area} = 0.7854 \, D^2$$
$$= 0.7854 \, (0.125 \text{ in.})^2$$
$$= 0.7854 \, (0.0156 \text{ in.}^2)$$
$$= 0.0123 \text{ in.}^2$$

9.1.16.1 Single Shear

An example of single shearing force is the action of a pair of scissor blades. In an assembly, two plates being held or a flywheel on a crankshaft would exert single shear stress. Single shear can be estimated by

$$\text{Shear force} = \text{Single shear strength (shear area)}$$
$$\text{Shear force} = 48{,}000 \text{ lb/in.}^2 \, (0.0123 \text{ in.}^2)$$
$$\text{Shear force} = 590 \text{ lb}$$

9.1.16.2 Double Shear

Many shear fasteners pin through two shear planes. An example would be a shaft and collar, pinned across their diameters:

$$\text{Double shear force} = 2 \, (\text{Single shear strength}) \, (\text{Area})$$
$$= 2 \, (48{,}000) \, (0.0123)$$
$$\text{Double shear force} = 1180 \text{ lb}$$

Figure 9.13 illustrates the shear mechanics.

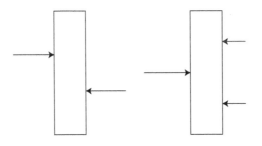

FIGURE 9.13
Single and double shear.

9.1.17 Calculating Torque–Tension Operating Charts

Step #1	Obtain equipment to measure torque applied
	Examples, torque wrench or torque transducer
Step #2	Obtain clamp load measuring equipment
	Examples, load cell or tensile tester
Step #3	Select a target clamp load
Step #4	Divide the load into five equal steps
Step #5	Apply torque until each step is obtained and record torque
Step #6	Plot the realitionship between torque input and clamp load output
Step #7	The plot gives an estimate for the torque–tension relationship for the conditions tested

Figure 9.14 illustrates the plot.

9.1.18 Estimating Assembly Efficiency

It is of value to know what percentage of assembly material strength is traded for the addition of an assembly member and the resulting assembled surface. One way to evaluate this is to find the strength of the fastened surface and compare it by ratio to the same surface if it were a solid piece. For example, two 1/4 in. square steel sections are fastened on end with a #10-32 screw made from the same steel raw material. We will say that the yield strength is 45,000 psi. The screw has a thread stress area of 0.2 in.[2]

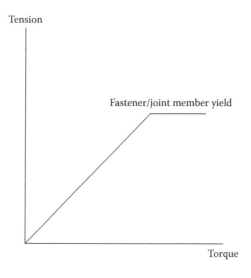

FIGURE 9.14
Torque–tension chart.

If our assembly is designed so that the screw thread is the weakest area, we would estimate joint strength as

$$\text{Joint strength} = \text{Area (Material yield strength)}$$
$$= 0.02 \text{ in.}^2(45{,}000 \text{ lb/in.}^2)$$
$$= 900 \text{ lb}$$

Our assembly has a yield limit of 9000 lb. If we now evaluate the joint as a solid member, we find:

$$\text{Base line strength} = \text{Area (Material strength)}$$
$$= (1/4 \text{ in.})(1/4 \text{ in.})(45{,}000 \text{ lb/in.}^2)$$
$$= (0.250 \text{ in.})(0.250 \text{ in.})(45{,}000 \text{ lb/in.}^2)$$
$$= (0.0625 \text{ in.}^2)(45{,}000 \text{ lb/in.}^2)$$
$$= 2812.5 \text{ lb}$$

Making a ratio of the fastened strength to the baseline strength, we find:

Fastener strength/Baseline strength
900 lb/2812.5 lb
0.32 or 32% joint efficiency

This is the efficiency cost of this joint interface. It could be improved by increasing fastener strength. Figure 9.15 illustrates the joint efficiency model.

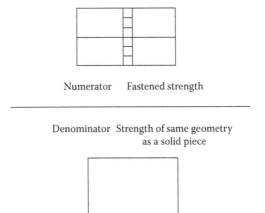

Numerator   Fastened strength

Denominator Strength of same geometry
as a solid piece

FIGURE 9.15
Joint efficiency model.

9.2 Work from the Workshop

In the fastener workshops mentioned earlier, a set of solved problems formed the basis for instruction. These were in inch system units, with a section set aside to discuss metric units. Following are those problems from my notes, first as originally presented in those workshops, and again in metric units with some notes which the reader might find useful. It was my observation during those workshops that a high percentage of students were mathematically capable if not always confident in their ability in this important fastener application area. I attributed this to lack of practice opportunity. At either tail of this normal distribution of numeric ability are those who have strong mathematics skills. These are often the result of good natural ability, a well-grounded educational background, and practice on real-world applications on a consistent basis. These workshop participants are a presenter's dream. The challenge when presenting is to not focus on these individuals and their questions to the detriment of others, whose math skills participating in the workshop with less formidable numeric ability. At the other end of the math skill's distribution can be found the math adverse. It is important that even those with limited skills, and confidence, be encouraged to try simple problems such as those that follow in the pursuit of good fastener fundamental skills. An analogy I would use is that someone learning to read would learn the alphabet. A musical instrument skill would involve knowledge and practice of scales. For fastener application work at even the most entry level position, good numeric skills are a useful and practical step worth taking. Good faith practice can lead to greater skill and confidence. For the instructor, presenting to the mean of the math skill of any given workshop group while engaging both tails of the skills distribution can be a challenge. Meeting it skillfully can provide positive outcomes for achieving the fastener trainings of both the individual participants and the workshop as a whole.

Fastener fundamentals, math examples, as presented:

Areas:

$$\text{Area} = 3.1416 \, d^2/4$$
$$= 0.7854 \, d^2$$

Example: A diameter of 1/2 in.

$$\text{Area} = 3.1416 \, (0.5")^2$$
$$= 0.1964 \text{ in.}^2$$

Volume:

$$\text{Volume} = \text{Area} \times \text{Height}$$
$$= [3.1416(d)^2/4] \times h$$

Example: A diameter of 1/2 in. and a height of 3/4 in.

$$\text{Volume} = [3.1416(0.5\text{''})^2/4] \times 0.750\text{''}$$
$$= 0.1964 \text{ in.}^2 \times 0.750\text{''}$$
$$= 0.147 \text{ in.}^3$$

Density: To find the weight of a part, we need to know the density of its material. We usually find this from its raw material properties. Density is the weight per cubic inch of its material.

The density of steel is approximately 0.28 lb/in.^{3}. Brass and stainless have similar raw material densities for fastener weight calculation purposes. Aluminum's density is about 0.11 lb/in.^{3}.

Weight: Let us calculate the weight of the part from the previous example, using steel as our raw material:

$$\text{Weight} = \text{Volume} \times \text{Density}$$

Or 0.147 in^3 × 0.28 lb/in.^{3} = 0.0412 lb with the in.^{3}'s cancelling out, leaving a result with units in lbs.

Note this cancelling out of units and a result in the expected units is a good proof when practicing fastener application engineering calculations.

For pounds per thousand parts, we multiply our result by 1000:

0.0412 lbs × 1000 = 41.16 or rounding to 41.2 pounds per thousand parts.

Part Cost Calculation: If our raw material steel costs $0.45 per pound, raw material cost would be:

$$41.2 \text{ lbs} \times \$0.45/\text{lb} = \$18.52 \text{ per thousand parts}$$

Note the lb units cancelling out yielding a result in dollars. In my shop experience, a scrap factor would be added to provide for wire coil ends, set up parts, and other nonsaleable uses of the production processes. A scrap factor of 5% would be typically budgeted and monthly scraps rates under 2% are certainly achievable. Let us calculate the volume, weight, and raw material cost of the following sample pin using steel at $0.45 per pound.

To find the volume, we will divide the part into two volumes, the head and the shank, and then add them. We will identify the head as section I and the shank as section II. The pin zones are cylindrical with head dimensions of diameter 0.375'' and head height 0.125'', the shank diameter 0.250'', and the shank length 0.625''.

$$\text{Volume I: } 3.1416(0.375 \text{ in.})^2 \, 0.125 \text{ in.}/4 = 0.0138 \text{ in.}^3$$

$$\text{Volume II: } 3.1416(0.250 \text{ in.})^2 \, 0.625 \text{ in.}/4 = 0.0307 \text{ in.}^3$$

$$\text{Part Volume} = \text{Section I Volume} + \text{Section II Volume}$$

$$= 0.0138 \text{ in.}^3 + 0.0445 \text{ in.}^3 = 0.0445 \text{ in.}^3.$$

Note that in this calculation, there is more than twice the volume in the shank due to its length than the head, which is larger in diameter. Multiplying the part volume by its raw material density:

$$0.0445 \text{ in.}^3 \times 0.28 \text{ lb/in.}^3 = 0.0125 \text{ lb}$$

Multiplying by 1000 to obtain the weight per thousand parts or by 100 to obtain weight per hundred parts: $0.0125 \text{ lb} \times 1000 = 12.46 \text{ lb}$, $0.0125 \text{ lb} \times 100 = 1.27 \text{ lb}$.

Multiplying by \$0.45/lb if that is our raw material cost and again 1.05 for a 5% scrap rate gives manufactured raw material cost for of headed pin fastener:

$$12.46 \text{ lb} \times \$0.45/\text{lb} \times 1.05 = \$5.89 \text{ per thousand pins}$$

9.2.1 Calculating Heading Upset Volumes and Diameter Ratios

A cold header manufacturing fastener blanks can "head" or upset approximately 2 1/4 "diameters" of raw material per blow. A conventional 2 blow cold header can upset $2 \times 2 \ 1/4$ or 4 1/2 diameters of material.
Example:

$$\text{Wire diameter} = 1/8\text{th of an inch} = 0.125''$$

Therefore one diameter of this raw material is 0.125" in length. 4 1/2 diameters is $4.5 \times 0.125'' = 0.563.''$
Heading steps would be

Cone blow

2nd or finishing blow

If the shank length is 1/2 in., a total material or cut-off length would be the upset length plus the shank length: $0.563'' + 0.500'' = 1.063.''$
The shaded zone is the amount of material upset and should not normally exceed the cold heading machine's limits.
Load:
A force applied to the structure and its fasteners, measured in pounds.
Stress:
Load applied over area.
Example:
Let us apply a tensile (pull-type) load in the area calculated previously at the beginning of these exercises. For ease of calculation, let us use a load of 1960 pounds.
To find the tensile stress, we divide the load by the area:

$$\text{Area} = 0.196 \text{ in.}^2 \text{ (from the preceding example)}$$

$$\text{Stress} = \text{Load/Area} = 1960 \text{ lbs.}/0.196 \text{ in.}^2$$

$$= 10,000 \text{ lbs/in}^2 \text{ also expresses as } 10,000 \text{ psi}$$

Torque:

Torque is a force that causes rotation. We apply torque to a wrench when we tighten a screw or bolt.

The simplified and not exact formula to find blot or screw torque is as follows:

$$\text{Torque} = \text{Friction} \times \text{Thread Nominal Diameter} \times \text{Clamp Load}$$

$$\text{Or } T = K \times D \times P$$

where torque is in inch-pounds, in-lb, nominal diameter is in inches, in., and load is in pounds, lb. Friction is a coefficient; no units are used with it. I have also explained it as an index that represents the friction in the fastening.

EXAMPLE #1

We apply 480 inch-pounds (or 40ft-lb × 12 in./ft) to a 1/2 in. diameter bolt. How much clamp load is developed?

Solution

$$T = K \times D \times P$$

$$480 \text{ in. lb} = 0.25(0.5 \text{ in.})P$$

$$P = 480 \text{ in. lb}/0.25(0.5 \text{ in.}) = 3840 \text{ lbs}$$

EXAMPLE #2

How much torque is required if 6000 lbs of clamp load is called for by the design and assembly instructions?

Solution

$$T = K \times D \times P$$

$$= 0.25(0.5 \text{ in.})(6000 \text{ lb})$$

$$= 750 \text{ in.-lb}$$

The preceding problems are from the notes used to present to many years of Fastener Fundamentals Workshop Attendees. I always mentioned that there are three fundamental equations that can help understand a wide range of the applications engineering problem encountered during normal practice. We presented the first two above: $S = P/A$ and $T = K \times D \times P$. The third we will add now:

$E = F \times L/A \times E$. This is an equation for elongation of a screw, bolt, or threaded stud. Listing them again:

$S = P/A$, the relationship quantifying fastener area, applied load, and resulting stress

$T = K \times D \times P$, the relationship linking tightening torque and clamp
load. It is an input torque, output assembly clamp load tool.

And $e = F \times L/A \times E$ provides a calculation of elongation under load. Let us
solve two problems with it.

E is elongation in the gripped area in inches and F is the load; we will use
the previous problem's 3840 lbs.

A is the area. We will use the preceding problem's 0.196 in.^{2} and E is the
modulus of elasticity of the fastener material. It can be considered that its rate
of strength would be below the yield point and is 30,000,000 lb/in^2.

We will use a grip length of 1.00 in.

For steel:

EXAMPLE

$$e = F \times L/A \times E \; e = 3840 \text{ lb} \times 1.00 \text{ in.}/0.196 \text{ in.}^2 \times 30,000,000 \text{ lb/in.}^2$$

$$e = 3840 \text{ lb-in.}/5,880,000 \text{ lb}$$

$$e = 0.0007 \text{ in.}$$

Our inch-long gripped length will stretch 0.0007 in. under the 3840 lb
applied load. If the grip length had been 2.00 in., it would have been 0.014 in.
Let us change the application slightly. We will use a 1/4-20 UNC-3A screw
with the 3840 lb load, 1.00 grip length (threaded), and manufactured from
the same steel with the same modulus of elasticity (E):

From the thread standards, the thread stress area for our 1/4-14 UNC-3A
Screw Thread is 0.0318 in.^{2}

$$e = F \times L/A \times E = 3840 \text{ lb} \times 1.00 \text{ in.}/0.0318 \text{ in.}^2 \times 30,000,000 \text{ lb/in.}^2$$

$$e = 3840 \text{ lb-in.}/954,000 \text{ lb}$$

$$e = 0.0040 \text{ in.}$$

As a rule of thumb, steel-threaded fasteners with yield strengths greater
than 100,000 lbs per square inch stretch mostly elastically slightly less than
0.004″ with an applied load developing stresses of 100,000 lb/in.^{2} for every
inch of grip length. And these three equations can be powerful tools in
understanding the static load application performance of many mechanical
fastening applications. As with any powerful tool, it should be used respon-
sibly and knowledgeably, backed by real-world testing, and tried out in the
fastener laboratory. Let us now solve these same types of problems using
metric units.

Areas: Area $= 3.1416 \, d^2/4$.

Example:

Area of a 12-mm diameter fastener grip diameter:

$$\text{Area} = 3.1416 \, (12 \text{ mm})^2/4$$

$$= 3.1416 \, (144 \text{ mm}^2)/4$$

$$= 113.097 \text{ mm}^2 \text{ or } 113.1 \text{ mm}^2$$

$$\text{Volume} = \text{Area} \times \text{Height}$$

We will use our 12-mm diameter and a grip length of 18 mm:

$$\text{Volume} = 3.1416 \, (12 \text{ mm})^2/4 \times 18 \text{ mm}$$

$$\text{Volume} = 2035.76 \text{ mm}^3 \text{ or } 203.58 \text{ cm}^3$$

Density:

In metric units, the physical properties are the same. The units used will be different. In the article section, I have placed an article titled *Universal Nature of Technology*. It was written at a time the company I was with was cold heading, rolling, heat treating, and plating tens of millions of small metric screws. The point I realized then and emphasize here is that the technology of fastener application is universal. The units from metric to inch system may be different, but the fastener science is the same. In calculating metric volume, we wrote the volume in cubic centimeters (cm^3). Density in metric units I have found to be expressed as grams per cubic millimeters or grams/cm^3. For our problem mentioned earlier,

Density: For steel, density is 7.75 g/cm^3. For aluminum it is 3.05 g/cm^3.

Weight:

Multiply metric unit volume by metric unit density: 203.58 cm^3 × 7.75 g/cm^3 = 1577.75 g.

Converting to kilograms we obtain: 1577.75 g/1000 g/kg = 1.58 kg.

We could obtain our raw material cost by multiplying by the cost per pound in euros, yuan, or whatever our desired currency and applying a suitable scrap factor.

One tool I have found to be of help with metric units is to list the inch and metric fastener strengths together as follows:

	Tensile Strength	Inch	Metric
Grade 5	120,000 psi	Grade 8.8	800 MPa
Grade 8	150,000 psi	Grade 10.9	1000 MPa
Socket cap screws, alloy steel	180,000 psi	Grade 2.0	1200 MPa

Making an index between two known values: 1800,000 psi to 1200 MPA Ratio = 1:150

So if I have a tensile strength of 100,000 lb/in.², it is approximately equivalent to 667 MPa.

Let us take a look at our three fundamental equations in metric units.

$S = P/A$ It is important for the units to be consistent. Stress will be in pascals, or megapascals if sufficiently large. Pascals are equal to newtons per square meter. Our force P will be in newtons and we will convert from square millimeters to square meters. In this way, the units are consistent and will cancel out during our algebraic calculations, yielding a solution in the proper metric units.

Let us take a load P of 18,000 N and apply it on a circular area of 113.1 mm², which converts to 0.0001 m² in four decimal places. I keep my calculator display set at four places. It calculates to more places and displays the rounded value. Our stress in newtons per square meter is 159,159.57, which is equal to 159 MPa.

For torque and tension, we will perform a similar operation to coordinate unit. Torque will be in new meters, diameter in meters, and load in newtons:

$$T = K \times D \times P$$

$$T = 0.2 \times 0.012 \text{ m} \times 18,000 \text{ N}$$

$$T = 0.2 \times 216 \text{ N m}$$

$$T = 43.2 \text{ N m}$$

And finally for elongation:

$$e = F \times L/A \times E$$

$$= 18,000 \text{ N} \times 0.0254 \text{ m}/0.0001 \text{ m}^2 \times 206,910 \text{ N/m}^2$$

$$= 0.021 \text{ m}$$

In the way of disclosure, I am definitely a mechanical man in an increasingly electronic age. That said, my first numeric language is the inch system, and I come to the metric system as a second language in which I am less than fluent. I hope the above combination of inch and metric examples are of help in understanding fasteners and assemblies. Following will be a series of case problem from my real-world experience as a fastener applications engineer. Most were dimensioned in inch system units. I will toggle from inch to metric units for the reader. The first is a favorite example and one used in numerous fastener fundamental workshops.

9.3 D & D 100

As presented in the article "A Socket Screw Remembered" in the appendix, the team which set as its goal the fix for sockets screws which were breaking in the field is a useful teaching example of fastener problem solving of a collaborative nature. Often the first information concerning a fastener application problem arrives in the form of a call to the office. That was the case with some large-diameter socket screws that my employer had sold to a distributor who supplied them to a Carolina-based manufacturer of a logging picking crane which was mounted on the back of flatbed trucks. The call came in the form of a distributor's sales representative. The first instructional point is that the easy conclusion to draw is that since the screws had broken, something is wrong with them. Jumping to conclusions is easy. A professional approach is usually the best: assume nothing; test and verify everything. In this application, that meant analyzing the fracture plane and a dimensional and metallurgical examination of the parts to the extent possible. The screw satisfied the dimensional and physical requirements per ASME and ASTM standards. The fracture plane was typical of metallurgical fatigue. This was communicated to the distributor and the end user. A field trip to further examine the application was arranged. With me went a newly acquired load cell and strain washer for the diameter fasteners in question. It had been acquired to demonstrate the torque–tension relationship to students in my fastener fundamentals workshops. Another teaching point is that I went to the site with all of the confidence that often only youth and a little success can bring. To my surprise, at the site was a team of very experienced engineers who, while not having my exact fastener knowledge skill set, were much more highly trained and equipped in machine design and dynamic force and reaction measurement. The important point here is that once acquainted, we quickly made the problem the boss. Its solution, regardless of where the solution took us, became the goal. I will stress again that it is most useful in fastener problem solving to make the problem the boss. There are no vested interests or egos to guard or protect. Also, adversarial I am right, you are wrong attitude should be dispensed with from the start. In the chapter, you can read the solution. The points are: Do your homework without jumping to conclusions, form a collaborative team, if possible, and make the problem the boss, and with a goal of the maximum benefit feasible solution.

References

1. Jim Speck, P.E., *American Fastener Journal*, Collected article problems and solutions.
2. SPEC Engineering, selected case problem work.

Appendix A: Fastening, Joining, and Assembly Glossary

Area

    Thread Stress Area: This is a theoretical area in a screw thread, which carries the assembly service and preloads. It is a circular area with a diameter of about the same size as the thread's pitch diameter. It is also very close to the roll thread diameter used for the fastener blank.

    Bearing area: This is the surface of a fastener, which seats against the assembled part and provides the clamping forces to hold the assembly together.

    Fillet area: The transition area from shank to bearing area is known as the fillet. These can be areas of increased stress and as such are shaped with radii or similar increased area features to carry these loads better.

Bearing The application of forces to clamp an assembly together. The bearing surfaces of a fastener distribute the fastening loads developed by the fastener over a mating surface on the assembly. As bearing areas increase for a given load, bearing stress decreases.

Belleville Washer An engineered spring action washer with a truncated cone geometry. When compressed by the clamping action of a tension fastener, it exerts a reaction force in an efficient manner due to its geometry and material strength.

Clamping Clamping is a basic action of a mechanical fastener. It is a compressing action against the assembly components. Clamping holds the components together.

Dog Point This is a feature manufactured onto the end of a fastener, often threaded, which allows it to better align itself into an assembly site. For example, a 2 mm dog point diameter, which is 1 mm long, could ease the assembly of a 3 mm threaded fastener into a tapped hole that is difficult to reach.

Fastener Any device that helps hold things together. A fastener can be a separate component or a feature on an assembly member. A fastener has three functions that should be addressed: strength, reusability, and appearance.

Fatigue Failure of a fastener after some period of time in an assembly. Often it can be from a small load, frequently applied. Metal fatigue can be seen in the failure of a metal paper clip after repeated back-and-forth bending. This is fatigue failure.

Fillet A radius or direction changing feature which connects fastener sections, such as heads and threads.

Gib Screw A set screw or any fastener used in compression to adjust clearance such as in machine tool sliding members. Gib screws often have plastic or brass inserts at their points.

Grip The length of a fastener that is squeezing the assembly members together. A tension fastener can have a designated grip length, which is sometimes an unthreaded section at full body diameter. The grip can also be the thickness of the assembly member actually being clamped.

Half Angle The angle formed by the triangle of a fastener's thread when seen in cross-section. For standard ANSI inch or metric threads, the half angle would be 30°. For special thread-forming threads or special threads such as a buttress or acme thread form, the half angles can vary. Thread angle also is a factor in thread gage measurement in combination with thread diameters, taper, and helix angle.

Helix The spiral form of a thread wrapping around the cylinder of the shank. In circular form, it is the basic mechanical inclined plane that supplies the wedging action to clamp assemblies together.

Indented A shallow cavity in the tops of cold-headed fasteners formed in cold-headed fastener blanks, which reduces the volume of upset material. By using an indent, a slightly larger head diameter, washer feature, or wrenching height can be cold headed.

"J" Thread A special increased radius thread form for high-strength applications, often used in roll threading after hardening.

Keps Nut A hex nut with an integral, free-spinning captive washer. It reduces parts count while still providing a free-spinning washer that can also provide some vibration resistance.

Locking Feature Most internal and external threads are "free turning." Negligible amounts of torque are needed to assemble and disassemble them. Even with sufficient preload, and more often without it, the vibration in some assemblies can cause the fasteners to "walk out" of their clamped positions. A locking feature on the external threads, the internal threads, between them or under the bearing area can provide vibration resistance. These locking features are available in many designs.

Multiple Thread Start The edge of thread at the very point of a bolt or screw is its starting thread. While one thread start and a single right-hand thread are typical, 2, 3 or more thread starts can lead onto a similar number of threads, which increase the advance per fastener rotation.

Nut An internally threaded, normally loose hardware component. Often thought of as a hex or square fastener, nut plates with multiple internal threaded holes as well as other shaped nut members are used.

Oval A curved, crown-shaped geometry on the face of some countersunk head screws as well as the points of some set screws.

Pilot An unthreaded section of a threaded fastener, which helps ease its location in the site during assembly.

Pinch Point A diamond-shaped point on a screw blank.

Preload A load placed on a fastener during installation.

Radius A blend from one fastener feature to another:

> **Fillet radius:** The usually curved section joining a fastener shank to its head at the bearing area. This is usually a high-stress area for a headed and threaded tension fastener. With a good fillet, these stresses are spread over an increased area and thereby reduced. Curves, angles such as 15° leading into a curved surface, and elliptical forms can all work and be manufactured. Two things to watch for are fillets that are so large in diameter that they interfere with the site's hole wall or chamfer and fillets that are sharp due to the manufactured surface grinding of the face of the heading die without relieving the edge later properly.

> **Thread root radius:** At the juncture of the pressure and nonpressure flanks of threads is a change in direction into a root flat or radius. This can also be a section of stress concentration. Proper radii in this area reduce the stresses without causing the thread minor and pitch diameters (for external threads) and major and pitch diameters (for internal threads) to go out of tolerance.

> **Thread runout radius:** The thread runout is the blend of incomplete last thread to either the fillet or unthreaded fastener shank. It is usually roll threaded or chased onto the fastener during manufacturing. If it is sharp, such as from the top surface of a thread rolling die that has been reground and the ground edge not properly broken, the thread runout can be a high-stress concentration area in the assembly.

Root The feature at the smallest diameter on an external thread and the largest diameter in an internal one. On threaded joints where performance is critical, attention should be paid to thread roots since microcracks and stress risers in these areas can be areas of performance problems. Techniques such as photomicrographic inspection, liquid penetration, and magnetic particle inspection can highlight potential thread root trouble spots.

Socket A hex, spline, six-lobed, or similar drive cavity formed into the head of tension fasteners such as socket, pan, fillister, flat, truss, oval, binding, or similar tension fasteners and the drive end of compression fasteners such as set screws.

Truss A head form designed to span a larger clamped area than a pan or similar head style.

Undercut A feature formed into the underside of a tension fastener head to allow closer threading to the head on very short screws as well as to provide a circular internal groove for a sealing feature such as an O-ring.

Wobble The characteristic of a cross-recess drive, which allows it to either stick firmly to or be loose on a drive matching driver bit. A wobble gage is available from gage suppliers. Another quick check at manufacturing is the fit of the cold-headed cross recess blanks on a new, unused punch from the same tooling lot being used. Good blanks will stick on the punch. Blanks with a wobble condition from heading with a worn punch will fall off the punch.

Appendix B: Ingenious Fasteners and Assemblies

We all start with the same laws of physics and economic market forces. We are also all held by time constraints on our work lives. Within these boundaries, fastening work advances to keep pace with the product development, which makes all of our lives more satisfying. Advances in fastening products and techniques are marked by gradual improvements and punctuated with leaps in fastening progress from new ways of holding components together.

In viewing the fasteners and assemblies in this appendix, keep in mind that they were championed by people who may have had to break new ground or face associates who pushed them in other, less efficient fastening directions. Perhaps their efforts were met with a "that can't be done" attitude rather than "can do" support. For every one of these, there are perhaps hundreds who tried a similar innovative fastening approach and were perhaps not as successful.

The companies and individuals who did persevere put their companies, their products, and themselves in a position to reap the commercial rewards that an innovative approach to fastening can offer, which range from the early American house construction approach of using post and beam, through early industrial assembly ideas such as interchangeable parts, to simple, easily assembled automobiles that many could afford, to the computers, peripherals, and miniaturized telecommunications equipment in our homes and workplaces today.

It is my hope that in considering these few select ingenious fastenings, the readers can then approach their important assembly work with fresh fastening ideas and creative assembly energy and that they can have a view not just to what has been tried and found true, but to ideas that can be inspired and built upon using their own insights, abilities, and assembly situations. Fastening is both an industrial science and a commercial art. When one individual champions a new and better way of fastening, we all benefit from their good work.

B.1 Eyeglass Hinge Screw

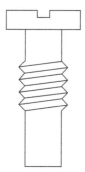

The hinge screw solved a frequent problem with the assembly of eyeglass temple pieces, which fit over your ears, with the front frame containing two lenses and with the screw rotating out of the internal thread with repeated movement of the temple pieces as they are used. The center section thread is a self-threading form for a line-to-line fit into the unthreaded site hole in the temple piece. The unthreaded sections above and below it on the screw provide a smooth hinge pin function for easier rotation. The point, being unthreaded, also provides a piloting action for high-speed automatic assembly machine if needed.

A key in many ingenious fastenings is the ability to meet a range of assembly requirements by designing several fastening functions, such as hinge action, clamping, piloting, and clamping all in one fastener. They make for efficient assembly, a lower parts count, reduced assembly cost, and a higher-value, better-performing assembled product that also has improved market esthetics.

B.2 Collared Hinge Pin

Taking the hinge screw one step further, the collared hinge pin shown provides a simpler and more robust locating, holding, and hinging eyeglass frame and temple piece with a precision cold-headed part and a well-designed injection-molded frame and temple pieces. The collar is press fit into a molded hole in the temple piece, leaving the two ends protruding from either side of the hole. The frame has a ramped channel with a wedge-like angle leading to two holes for these hinge pin ends to snap into. The ends of the hinge pin are radiused. When the temple piece, with hinge pin installed, is pressed into the channel in the frame, the plastic frame is pried open slightly, allowing the length of the pin to just line up with the centerline of the two holes. At that point, the holes snap onto the hinge pin ends, providing an assembled eyeglass. The installation is rapid, low in cost, and very strong. Moreover, no metal parts are exposed, yet the metal hinge pin provides higher joint strength than if plastic pin ends were molded into the temple pieces. Reusability is low but not a prime requirement in this product's target market. And the innovative hinge design provides the customer with high joint integrity and good assembly appearance.

B.3 Rivet Screw

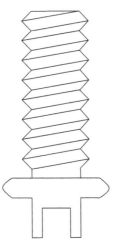

This design provides the solution for a good many applications. That is the need for a threaded attachment for subsequent assembly The top half of this headed part is formed as a semi-tubular rivet. The bottom half is a conventional machine screw. In application, the semi-tubular rivet feature of this one-piece fastener permits use of an economical rivet into the mating part. The riveting is quick and of adequate tensile and shear strength. The units do not need to be taken apart for service. A threaded nut is used to attach another member to the assembly. They do need to have a connector attached to the threaded end, as well, in the field. And in this case, the threads are used as a gripping surface for the connector, which is pressed onto, rather than threaded, to make the connection. The rivet screw functions as a rivet, as a threaded tension fastener and as a friction connection.

B.4 Fuel Pump Rivets

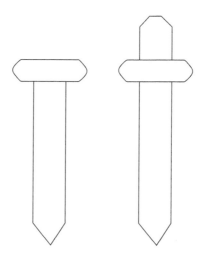

A fuel pump manufacturer was faced with the task of downsizing the assembly size of an in-tank fuel pump. Of course, a smaller assembly requires smaller fasteners. They also had to be made from a material impervious to petroleum-based fuels. The design shown is cold headed and roll formed from 300 series stainless steel. The roll-formed point is able to align several pump sections as it is pushed through the pump assembly. The point is then orbitally riveted. Two rivets, one of each type shown, are used. As the pump cap is assembled after the riveting step, the stem on the longer rivet provides a locating feature, so the inlet and outlet ports in the cover always line up with the corresponding ports on the pump. From the collar down, the rivets are identical, so common tooling can be used.

B.5 Self-Threading Screws

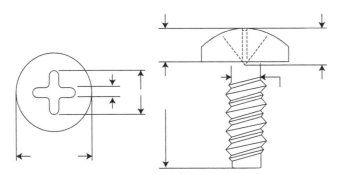

The ability to delete a tapping operation for internal threads and proceed directly to installing a screw into an unthreaded hole is the kind of fastener innovation that allows quantum leaps in assembly efficiency. Starting with wood screw threads, and advancing through lettered self-tappers such as A, AB, B, C, and F and on to proprietary engineered thread designs that modify pitch, thread including angles and half angles, combined, two alternate thread angles and rolled lobes and other diametral designs on their blanks, they all have advanced assembly-operations worldwide. Supported by the roll-threading die suppliers who provide the tools and technical support, any assembler can access the optimum design to achieve an efficient strip-to-drive ratio for specific fastening sites.

B.6 Automobile Lug Nut

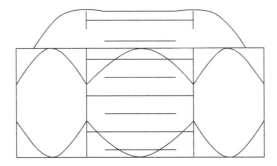

While it is ubiquitous and easy to overlook, the auto wheel lug nut is ingenious in its combination of features to bring high reliability, ease of use, and economical assembly to motorists everywhere. The lug nut is not just a hex nut carrying a tensile load. The curved, bull nose feature of the lug nut nests into the mating concavity stamped into the wheel to form a strong shear pin for carrying the high reversing shear stresses encountered as the car accelerates and decelerates. At the same time, the nut threads and curved bearing area resist the tensile forces as the wheel corners and forces are exerted on axis relatively parallel to the wheel lug/lug nut axis. The lug nut, along with the mating wheel lug and wheel fastening site concavity, combines the fastener functions of a shear fastener and a tensile fastener in one efficient fastening design.

B.7 Plastic Christmas Tree Fastener

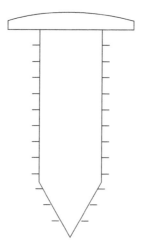

Fasteners are often developed to satisfy the changing nature of the materials they hold together. As durable goods makers switched from all sheet metal construction for cars and household appliances, to thermoplastic panels and components, the plastic snap action figure was developed to meet the application requirements and has met with great acceptance.

Its appearance can be matched to the color and texture of the surrounding assembly. It offers superior assembly ease together with adequate holding power, and it is easy to remove and replace as the application requires. It may not have the strength of a metal fastener, but its all-round performance of strength, reusability, and appearance is hard to beat in many plastic component fastening applications.

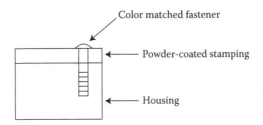

B.8 Anaerobic Adhesive

Developed as an improvement in "glue," these engineered adhesives were first scoffed at by skeptics who were convinced adhesives had no place on the industrial fastening site. After a false start during which they could not be "given away," anaerobic adhesives that cure in the absence of air, as during the tightening of engaged screw threads, gained the confidence of engineers, designers, and consumers. Anaerobic adhesives increased assembly reliability and efficiency.

Providing an all-inclusive list of ingenious fasteners would be a Herculean task. Too much innovation has been, and continues to be, applied in the fastening industry. It would be a serious omission, however, to exclude the development and commercial refinement of anaerobic adhesives.

Curing in the absence of air, these high-strength engineered adhesives have solved assembly problems in fastening applications ranging from vibration resistance in the clearance between clamped threads and locking bearings into their races to being the sole fastening mechanism in the application.

Although they have low reusability, they can have high value in terms of appearance and strength and have improved dramatically the options of assembly designers everywhere.

B.9 Theft-Resistant Fasteners

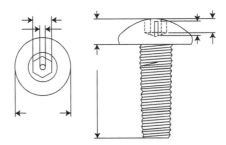

Many techniques and designs have been tried over the years to thwart would-be thieves from loosening an assembly's fasteners and making off with the product. In the design shown here, two different fastener features are added to deter crime. The fastener drive has an integral pin cold headed in the socket, which extends up to the face of the head. Any wrench used to loosen the screw would have to have a mating hole. Since the wrenches are normally heat treated, drilling a hole would require some additional work. The screw is then installed in a counterbored hole; some gripping tools cannot be used on the outer diameter of the head. Another ingenious, yet simple, fastener feature is the use of left-handed rather than the conventional right-handed threads on the fastener and in the tapped hole. The screws are installed by turning counterclockwise. A vandal, on defeating the tamper-resistant drive, would then be trying to tighten rather than loosen the screw by applying torque in the usual manner.

B.10 Multiple Start Threads

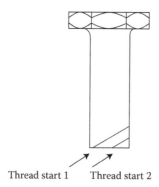

Thread start 1   Thread start 2

The name of the clever person who first learned to combine two or more threads on a common fastener shank has no doubt been lost in fastening history. But the use of multiple thread starts can provide exceptionally rapid assembly plus the reusability of threaded fastening in applications where conventional single lead threads would prove less efficient.

B.11 Roll-Formed Fasteners

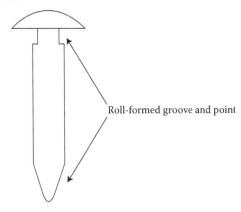

Roll-formed groove and point

Here is one fastening inventor whose name has not been lost. Roger Orlomoski developed the concept of using thread rolling machinery to produce grooves, collars, special points, and other geometries on fastener blanks that were previously only possible by machining. At his company, Rol-Flo Engineering, ingenuity of roll-forming special fastener shapes continues.

Appendix C: Some Frequent Fastener Questions and Answers

Question:

When is an application a good one for threaded fasteners?

Answer:

Threaded fasteners are most useful when there is a need for reusability that has some frequency. For example, changing the wheels on an automobile makes good application of machine screw threads. Threads would not be as useful on the exterior trim molding, which may be held on with snap-on clips or adhesive. When reusability is a possibility, screw threads are often a good candidate solution for the application. When the frequency of reuse is uncertain or infrequent enough to not require the high reusability of threaded fasteners, other fastener designs deserve evaluation and, if appropriate, testing.

Question:

What is hydrogen embrittlement and how does it occur in fasteners?

Answer:

Hydrogen embrittlement is a metallurgical phenomenon which makes steel fasteners brittle and easily fractured under load. It is the result of hydrogen accumulating in fasteners, usually during electroplating. An example would be an alloy steel tension fastener, through hardened to HRc 40 and cleaned or zinc plated in a plating process generating excessive amounts of hydrogen gas as a by-product. The hydrogen migrates to the highly stressed areas of the fastener. During tensioning, the molecules of hydrogen attach themselves to the grains of steel composed of atomic iron, carbon, manganese, and the other alloying elements, taking away the space that normally allows the fastener grains to flex under load and remain somewhat ductile. Instead, the grain boundaries develop cracks that grow quickly under load and result in the fastener fracturing, sometimes catastrophically.

Hydrogen embrittlement can be controlled by three steps. First, the cleaning and plating processes should be selected, controlled, and monitored. Electroless nickel is not known to be prone to hydrogen embrittling, while zinc plating is. The plating process should have adequate process controls for current density, which is a contributing factor to excessive hydrogen generation. Second, all heat-treated and electroplated fasteners should be baked at a minimum of 375°F, for at least 3 h in a temperature-controlled oven for 1 h of plating.

This baking for embrittlement relief is very effective in preventing the hydrogen buildup in plated fasteners, which causes hydrogen embrittlement. Process documentation can be used to verify that the embrittlement relief baking has been carried out. Finally, a sample from the plated lot of fasteners should be tested for hydrogen embrittlement using a simple stress durability test. This test loads the specimen fasteners to a high tensile load under controlled conditions, where they are held for at least 200 h. Embrittled fasteners will usually fail this test. The stress durability test can identify hydrogen embrittlement in a lot of plated fasteners. It has been observed that zinc-plated fasteners failing the stress durability test due to hydrogen embrittlement have been made functional by stripping the zinc plating, baking the fasteners at 750°F for 3 h and then replating with electroless nickel.

Question:

What causes fastener stripping?

Answer:

Threaded fasteners strip because the load on the pressure flanks of the internal and/or external threads became larger than the yield strength of the material from which they were manufactured. It is important in these applications to realize that the majority of the load is carried by the first few full threads closest to the bearing areas. Using sufficient area and fastener materials of high enough strength for the application helps prevent fastener stripping.

Question:

What causes fastener recess drive cam-out and reaming?

Answer:

Similar to the previous answer, the drive torque exceeds the drive area and its yield strength. To decrease the opportunity for drive failure, use well-maintained and fitting tools in fasteners made with drives to specification. Be aware of cutting corners during fastener driving by not setting tools fully on the fastener or at improper angles. In high-volume fastener applications, plan the tools, fasteners, and workplace to minimize the opportunities for improper driving practices.

Question:

How does plating and fastener finish affect fastener performance?

Answer:

Plating and general fastener finish control the amount of friction present during installation of the fasteners, as estimated by the "*K*" factor in the torque–tension equation. In general, a porous plating and a rough plating finish will require more torque to achieve the same preload as a finish which is smooth or less porous. Alternatively, at the same installed torque levels, the higher friction plating will be preloaded at a lower level. Corrosion resistance as measured by salt spray resistance or other indicators can also vary.

Question:

How does zinc plating differ from cadmium plating of fasteners?

Answer:

While both are metallic and applied to fasteners generally by electroplating processes, zinc plating has a rougher surface and a higher thread coefficient of friction. Cadmium plating is smoother and has a lower coefficient of friction. Both can be processed through chromic acid solutions that remove small amounts of the plating but smoothen the fastener's finish, close up porosity, and provide a different color if a die such as yellow, olive drab, or blue has been added to the chromate solution. The resulting finish is sometimes called by its color, such as zinc and yellow dichromate. It is thought that for equal plating thickness, cadmium provides superior corrosion resistance. This is true for marine environments, where salt is present in the air. In sulfur-rich environments such as found along a highway or in heavily industrialized areas, zinc plating actually has a slightly superior corrosion resistance. For this reason, it is beneficial to know a fastener's service environment.

Question:

What function do jam nuts serve?

Answer:

Jam nuts, which have a partial height of their full hex nut counterparts, are designed to be applied "in series" with a full hex nut. Their purpose is to draw some of the thread flank pressure off the full hex pressure flanks and onto the jam nut flanks. This results in more uniform thread pressure distribution than could be obtained with an extra thick nut. The jam screw is torqued to half of the torque of the full hex nut.

Question:

Is vibrational resistance affected by thread fit?

Answer:

In general, for free assembling machine screw threads, transverse vibrational resistance is more a function of preload and thread friction than thread fit. It is only when interference or special locking thread geometries are used in the application that vibrational resistance can be significantly increased.

Question:

Is tensile strength affected by elevated temperatures?

Answer:

Yes. As a fastener's service temperature rises, its tensile strength decreases. For fasteners that are exposed to temperatures near their tempering temperatures, retempering can occur if the original tempering temperature is exceeded. At temperatures below retempering, but well above room temperature, say 350–500°F, the phenomenon of creep occurs. Creep is the loss of strength with time at elevated temperature. If raised temperatures are part of the service environment, the creep characteristics and tempering temperature of the fastener and assembled component materials should be taken into consideration.

Question:

How can thread gaging quality disputes be resolved?

Answer:

In situations where two different calibrated machine screw gaging systems disagree on the acceptability of a machine screw-threaded fastener, it is good practice to review the gage setting standards and the master calibration gages used. In applications where one gage system is held to a tighter calibration tolerance, but both gages meet gage calibration standards, the operation of the subject fasteners in the intended service should be evaluated from an engineering perspective. Often, the difference in pitch diameter measurement being disputed is 0.0001–0.0002 in.

Question:

What is a common assembly mistake?

Answer:

Tightening torque is often used as a fastening performance measurement when the resulting tensile preload is of more importance. Since the tensile preload can vary depending on the amount of fastening friction as estimated by the "K" coefficient value, the actual preload obtained for a measured tightening torque could vary by a wide percentage. Also, any misalignment or bending could scrub off additional torque, with less of it going into elongating the tensile fastener and building preload.

Question:
What is work hardening and how does it affect fasteners?

Answer:
Work hardening is the metallurgical strengthening of a fastener raw material through cold heading, roll threading, and other metal-forming processes. It is relieved during heat treatment, but on non-heat-treatable fasteners such as 300 series stainless steel it can represent a significant gain in strength. Some high-strength heat-treated fasteners are thread rolled after heat treatment and have their fillets cold worked to take advantage of these work hardening processes.

Question:
Can a fastener be too strong?

Answer:
Yes. The important consideration is to match the fastener's strength in the areas needed, such as tensile strength, ductility, fatigue strength, impact strength, and creep strength, and select the one best suited to the intended service. A fastener which is high in tensile strength but lower in ductility, if this is more application critical, is not the best choice. It is best to view fastener strength as a curve not a straight line. The best fastener will maximize the area under the curve.

Question:
What is a fastener?

Answer:
Anything that holds things together.

Question:
When should washers be used?

Answer:
In some applications, a washer can increase the clamp-affected zone to bring a larger mass of the assembly member(s) into the fastening work. Washers can also reduce head bearing area friction and increase grip length. Joint faces and the parts count increase, however. The decision of whether to use washers has to be made by the assembly designer and engineer based on the specifics of the application. It can be valuable to make the decision on

the basis of how to accomplish the washer's functions in the assembly rather than considering the washer as a separate assembly part. The positive functions may then be achieved in the most efficient manner.

Question:

Are cold-headed fasteners always better than machined fasteners?

Answer:

No. The most efficient fastener manufacturing process depends on the design of the fastener, the material used, and the quantity to be purchased. Screw machining may be more efficient if smaller quantities or sharp features and tight tolerances are required. Cold heading may be more efficient if slightly more generous tolerances or larger volumes are anticipated. It can sometimes make economic and performance sense to start with a machined fastener and convert to cold heading as the assembly production ramps up. The engineering differences in the fasteners produced by these two different manufacturing processes should be carefully considered, however.

Question:

What are some sources for fastening information specific to my application?

Answer:

A good starting point is with your local fastener supplier in concert with the manufacturers who supply them. The local distributor can provide information on what fastening products are available, who makes them, and the names of factory technical support personnel. The local fastener distributor can open up the fastening information market to you. By making them a team member of your fastening project, you can take advantage of the full range of fastening information that can take your application in the direction of world-class fastening practice.

Appendix D: Article from American Fastener Journal

Reprinted from *American Fastener Journal,* July/August 1990

Design for Assembly:
Fastener Distributor's Problem or Opportunity

By Jim Speck, P.E.

As fastener manufacturers and distributors, it is important to carefully study an engineering technique which is becoming popular with design engineers. It is making its presence felt in the new products coming off of engineer's computer screens.

Design for Assembly, or DFA, as it is commonly called, gives fastener users a powerful tool to eliminate a large percentage of the 25% to 50% that assembly cost represents in the total manufacturing cost of their products. That's good news for users of this new technology. It's bad news for fastener consumption because one of the major objectives of DFA techniques is to reduce parts count. And fasteners of all types are being designed out. For example, every year one leading trade magazine presents an award for productivity through assembly technology to the country's outstanding example of assembly technology and thinking. A recent recipient won top honors for rigorously applying DFA methods in the development of a retail sales terminal. This winning design featured:

- 65% fewer suppliers;
- 75% less assembly time, with 100% fewer assembly tolls (it can now be hand assembled); and
- 85% fewer parts (100% fewer screws).

In fact the whole product has fewer parts than the previous one had screws. This is obviously a leading edge type product. It does however represent a trend which can have a positive or negative effect on your business depending on your approach.

If you take a "business as usual" approach and plan on low price, and good service to hold your customers, you may not lose accounts to competitors. But you will lose that business when the fasteners are designed out. Either way you lose the business.

On the contrary, if you build a case for Best Cost Assembly or BCA for their current or proposed product, you come out ahead in terms of total cost compared to a Design for Assembly effort; keeping the fastener business and enhancing your value to your customer. Because, with all its advantages, the one area where DFA is inefficient is in start up cost. It requires considerable engineering and computer time and money to first perform the Design for Assembly analysis and then tool up and manufacture the fewer, more expensive pieces. Unless a manufacturer is assured of high sales volume for their product, a good case can be made for simple for simple fastening as long as it is cost effective and efficient.

> # Design for Assembly, or DFA, as it is commonly called, gives fastener users a powerful tool to eliminate a large percentage of the 25% to 50% that assembly cost represents in the total manufacturing cost of their products.

And in this era of condensed design to market times and rapid technological advances in products, product's life cycles are being shortened. The designer's job is to lay out he highest value products possible at the lowest possible cost. If the product fills its intended need in the market, this is the route to greatest market share and profitability.

Designers and their companies are recognizing that using DFA or BCA can significantly reduce the cost of the products and that savings made at the design stage are good for the life of the products.

Most products can be analyzed for assembly efficiency by exploding them into individual parts, and making a diagrammatic analysis of the assembly line time and cost of each part.

Let's think for a moment about the ordinary tape rule. It is comprised of a case (containing a coiled tape rule) made up of two halves fastened at the corners. A belt clip is fastened to the back. The questions the designer will ask are:

1. Do the parts have to MOVE relative to each other?
2. Can they be of the same MATERIAL?
3. Do they have to be taken apart to be MAINTAINED?

These are the three M questions: MOVE, MATERIAL and MAINTENANCE.

A tree diagram can be made with individual parts on the horizontal axis and assemblies on the vertical axis as seen in the accompanying figure.

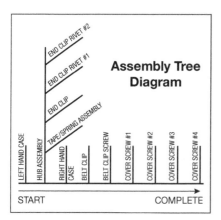

Assembly Tree Diagram

The tree diagram suggests part reduction and assembly efficiency increases. Two semi-tubular rivets might be replaced with one rivet and a tab stamped into the blade during manufac-

turing. This would limit transverse movement like the second rivet was doing.

The four screws can be replaced with two in conjunction with a lengthening of the case lip with posts and sockets molded into the new case halves. It may even be possible to design the case halves so that it can be used as either right or left rather than having a separate design for each of the case halves.

The designer's job is to lay out the highest value products possible at the lowest possible cost.

Rather than having a screw to hold on the belt clip, why not make a headed part to hold the clip, case half in the center and also serve as pivot and anchor for the tape/spring assembly? In fact, you might be able to eliminate the remaining corner screws.

Now, to do this, many engineering questions have to be answered. How many pounds (or newtons if we want to go metric to compete worldwide) of pull-out strength does the new center screw have versus the assembly with four corner screws?

What is the relative strength of the new versus the old belt clip attachment? And how much will the new parts cost?

Fastener distributors have traditionally been sharp at answering the latter question. But to compete and win in this changing Design for Assembly environment, they will also have to be prepared to respond to more of the former types of questions.

This is least cost assembly language and is value based rather than price based. And it could lead you to better long term relationships with your customers. ❏

Reprinted from *American Fastener Journal,* November/December 1991

Answering Customer Questions About Fastener Head and Shoulder Dimensions

By Jim Speck, P.E.

Buyers, purchasing agents and engineers are often interested in the standard configurations and dimensions of fasteners. Literature showing this information for standard types is useful to have on hand to handle the majority of these requests. But how do you respond to the request of "How much larger can we make this head?"

Well, one way to respond would be to fax the request to your best source(s) and push for a quick reply, then expedite it back to your customer—and hope they didn't get the answer from another source first. Or, a more efficient approach would be for you to be able to whip out your calculator and notes from your briefcase and provide the answer right at the customer's desk. With the right information and some practice, any good sales rep can do it. And working through the problems will give you a better feel for the cold heading process, its limits, and the capabilities of the fastener factories you work with, which in itself is a good deal.

In figuring possible head dimensions, you need to take the head diameter, calculate the area, multiply it by the head height to obtain the head volume. Let's work through four examples that follow.

Again the steps are (if head is close to cylindrically shaped, otherwise use the appropriate shape formula:

1. Multiply the head diameter by itself (squaring it)
2. Multiply the result by 0.7854 (pi divided by 4),
3. This gives head area, now multiply by head height. This gives head volume not counting any recesses, sockets, etc.

4. Subtract any recesses, sockets or cavities in the head

With head volume, you can now estimate diameters of wire in the proposed head. A manufacturer would use the fastener's roll thread diameter but the pitch diameter is just as useful for your customer calculations. They are also worth having in chart form in your portable fastener fact file.

Here is an example of calculating cut off length and diameters in the proposed head using the last volume we calculated and using a machine screw thread pitch diameter:

Head Volume from Example #4, Cross Recess Flat Head = 0.0011 in²

Shank Area = .7854 (.095 in)² = .0071 in²

Cut-Off Length in Head
Head Volume(in³)/Shank Area(in²) =
.0011 in³/.0071 in² = 0.1552 in²

Number of diameters in the head
.1552 in/.095 in = 1.63

The answer, length of pitch diameter wire in the head, is divided by the pitch diameter to give the number of diameters in the head. In this case the number of diameters would be 1.63. You can use 4–4.5 diameters for most of the standard cold heading raw materials if your sources are using conventional single die-two blow headers. If the number of diameters is larger than 4.5, the part may still be headable, but a header with more blows (2 die-3 blow, 3 die-3 blow, etc.) may be needed. You can advise your customer that the extra forming blows can add to the cost. And using the equations, it is possible to make trade-offs between head diameters and head height and perform "what if" upset diameter calculations.

If the threads are going to be thread forming or thread cutting, you can make an adjustment

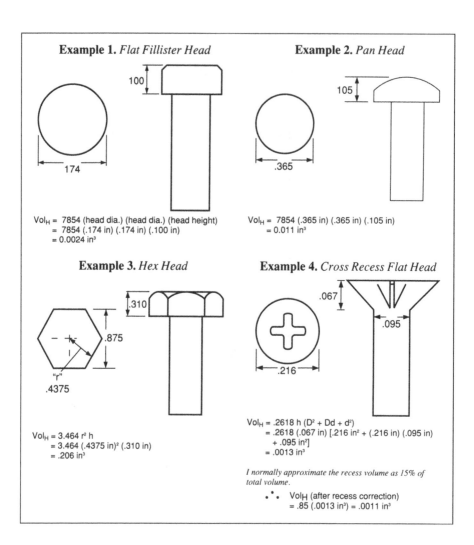

Example 1. *Flat Fillister Head*

100

174

Vol$_H$ = 7854 (head dia.) (head dia.) (head height)
  = 7854 (.174 in) (.174 in) (.100 in)
  = 0.0024 in³

Example 2. *Pan Head*

105

.365

Vol$_H$ = 7854 (.365 in) (.365 in) (.105 in)
  = 0.011 in³

Example 3. *Hex Head*

.310

.875

"r"
.4375

Vol$_H$ = 3.464 r² h
  = 3.464 (.4375 in)² (.310 in)
  = .206 in³

Example 4. *Cross Recess Flat Head*

.067

.095

.216

Vol$_H$ = .2618 h (D² + Dd + d²)
  = .2618 (.067 in) [.216 in² + (.216 in) (.095 in)
  + .095 in²]
  = .0013 in³

I normally approximate the recess volume as 15% of total volume.

∴ Vol$_H$ (after recess correction)
  = .85 (.0013 in³) = .0011 in³

of the required number of thousandths to the equivalent machine screw pitch diameter. Normally a spaced thread will be several thousandths smaller than its machine screw counterpart. If you practice doing the head diameter and upset volume calculations on some of the standard IFI head styles, you will be surprised at how proficient you can become in cranking through different special head dimensions. You can perform a professional service for your customers and prospects which is bound to create an image of someone who takes their profession seriously and who knows how to do their homework. ❏

Reprinted from *American Fastener Journal,* January/February 1992

Mechanical Advantage

By Jim Speck, P.E.

One of the main benefits of using metal fasteners is the mechanical advantage they give the user. But often I the rush of day-to-day business, we overlook some of the benefit those "mechanical advantages" afford our prospects and bring to our customers. And more overlooked are the skills and techniques available to fine tune these mechanical advantages.

When a professional race car driver or golf pro go to a new course, respectively, they take the time to really look at, to observe, all of the details of the new coarse. Only then can they best apply their skills to this new situation.

Assemblies are the same way. Each has its own set of details and features that need to be observed if we are to apply our skills in recommending the "Best Cost" assembly approach. Two of the skills we should study to master so that we offer the most competitive fasteners on both a performance and in-place cost basis are engineering mechanics and strength of materials.

Engineering mechanics is the analysis of forces applied to everyday mechanisms. For the fastener professional, engineering mechanics means taking the viewpoint that each force in the assembly can be viewed as a vector (or arrow) with the length being the amount of the force and the arrow's direction being its sense, in other words, the direction it pushes or pulls the part(s).

Strength of materials is just as its name implies; the strains a material develops under load, in response to the applied forces, and the stresses it develops in responding.

As a case study of the skills, let's consider three very similar #4 Type "B" screws tightened to 10 in-lb to assemble a plastic component. Let's vary the head style to give three different head bearing areas as shown in the figure and have a look at the resulting bearing stresses as follows:

#4-24 Type "B" Assembling Torque = 10 lb-in. Resulting clamping load = 400 lb. Compare the bearing stress for three head style/diameters (pan, binding, truss) and the first engaged thread pressure flank stress for major diameters at two different values.

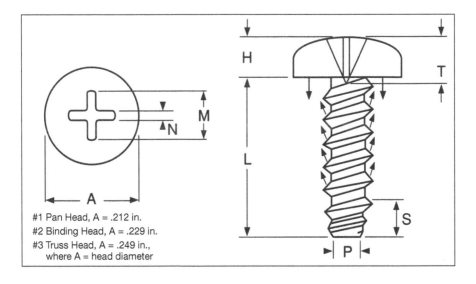

#1 Pan Head, A = .212 in.
#2 Binding Head, A = .229 in.
#3 Truss Head, A = .249 in.,
    where A = head diameter

#1 Pan Head

Bearing Area $= .7854 (.212 \text{ in} - .093 \text{ in})^2$
               $= .0111 \text{ in}^2$

Applied Load $= 400 \text{ lb}$

Bearing Stress $= 400\text{lb}/.0111 \text{ in}^2$
               $= 36,036 \text{ lb/in}^2$

#2 Binding Head

Bearing Area $= .7854 (.229 \text{ in} - .093 \text{in})^2$
               $= .0145 \text{ in}^2$

Applied Load $= 400 \text{ lb}$

Bearing Stress $= 400 \text{ lb}/.0145 \text{ in}^2$
               $= 27,586 \text{ lb/in}^2$

A 25% improvement over #1.

#3 Truss Head

Bearing Area $= .7854 (.249 \text{ in} - .093 \text{ in})^2$
               $= .0191 \text{ in}^2$
Applied Load $= 400 \text{ lb}$
Bearing Stress $= 400 \text{ lb}/.0191 \text{ in}^2$
               $= 20,942 \text{ lb/in}^2$

A 24.1% improvement over #2.

Now, let's compare one measure of thread stress, namely pressure flank stress, for the thread major diameter at two different sizes:

Take first fully engaged thread as carrying 30% of clamp load.

400 lb (0.3) = 120 lbs. Threads are 60° included angle.

30° 1/2 angle. Use cosine of 30° to find pressure area.

#1 Pressure flank for .110 in. major diameter and .084 in. minor diameter

Pressure Flank Area
$= .7854 [(.110\text{in})^2 - (.084\text{in})^2]/\text{cosine } 30°$
$= .7854 (.005 \text{ in}^2)/.866$
$= .0046 \text{ in}^2$

Pressure Flank Stress
$= 120 \text{ lbs}/.0046 \text{ in}^2$
$= 26,087 \text{ lb/in}^2$

#2 Pressure flank for .112 in. major diameter and .084 in. minor diameter

Pressure Flank Area
$= .7854 [(.112 \text{ in})^2 - (.084 \text{ in})^2]/\text{cosine } 30°$
$= .7854(.0055 \text{ in}^2)/.866$
$= .0050 \text{ in}^2$

Pressure Flank Stress
$= 120 \text{ lbs}/.0050 \text{ in}^2$
$= 24,000 \text{ lb/in}^2$

This represents an 8.7% improvement over the .110 major diameter, all other things being equal. It is attention to details like these which separate the well-engineered product from the competition. So go to school on the details and your business will reap dividends. ❑

Reprinted from *American Fastener Journal,* March/April 1992

Handling Special Fastener Applications

By Jim Speck, P.E.

Fastener applications can present today's fastener professionals with some tough questions. What type of fastener should a customer use for a given set of application requirements? What information, what questions need to be asked, to cross over the threshold of one of many possible suppliers to the better relationship of valued supplier? How can we be assured that our fastener represents the maximum cost efficiency and is the "Best Cost" solution?

The fasteners we supply must be capable of withstanding loads applied to them throughout their lives. But at the same time, over-specifying can wind up costing the customer more, raising their product costs and lowering profit.

The critical problem is to economically help the customer to fasten their product(s) in a method that meets the stringent requirements of safety, efficiency and long life.

Appropriate fastener selection is one of the key considerations in assuring assembly integrity, but it is often overlooked. To choose the best possible fastener for an application, we should consider these questions:

- What are the fastener's functions?
- What static and dynamic loads are on the fasteners?
- What types of fasteners are applicable?
- Should the fasteners be removable? How often will they be removed?
- Are the fasteners safe for the application?
- Are they economically efficient?

Determine the Functions

The first step in selecting fasteners is to determine what they will be expected to do in the assembled product. In most cases, their primary function is to hold parts together. But fasteners may act as locators, guides, pivots or in other roles.

Fasteners may also have an appearance function—a certain color or texture finish may be required, or perhaps they need to blend in or be flush with the fastening site surface.

Analyze the Loads

A listing of the fasteners' functions can serve as the base for load analysis. Fasteners often apply compressive force—in other words they restrain tensile loading. But they may resist shear, torque, bending or compressive forces. Loads can be divided into two categories: static and dynamics as shown in Figure 1.

Static loads determine the yield and tensile strengths required in the fasteners and in the fastened joint. Assuming impact is negligible, the yield strength of the fasteners is generally used as the basis for calculations.

Dynamics loading, on the other hand, is concerned with the fastener fatigue strength. This factor is a measure of fastener endurance limit. The fact that fatigue-propagation failure often has a higher loss function suggests that this design criterion merits serious attention.

The magnitude and frequency of applied load(s) along with their type (tensile, compressive, shear or combined) should be determined. Fatigue factors should be considered and the chosen fastener's endurance limit should be greater than the cyclical dynamic loading in the application.

Many applications aren't just static however. They are dynamically loaded, and we need to consider fatigue strength, and possibly impact strength.

Choose the Fastener Type

Fastening products vary from adhesives to zippers. If the components will never be disassembled, permanent fastening methods such as welding, brazing, soldering, adhesives or rivets should be considered. If the components are to be separated only very rarely, low-reusability systems might represent the "Best

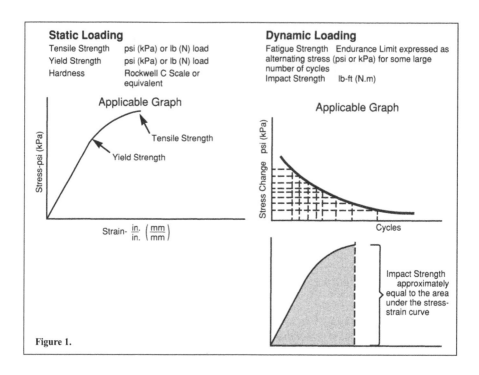

Static Loading

Tensile Strength	psi (kPa) or lb (N) load
Yield Strength	psi (kPa) or lb (N) load
Hardness	Rockwell C Scale or equivalent

Applicable Graph

Stress-psi (kPa)

Tensile Strength

Yield Strength

Strain- $\frac{in.}{in.}$ $\left(\frac{mm}{mm}\right)$

Dynamic Loading

Fatigue Strength   Endurance Limit expressed as alternating stress (psi or kPa) for some large number of cycles
Impact Strength   lb-ft (N.m)

Applicable Graph

Stress Change   psi (kPa)

Cycles

Impact Strength approximately equal to the area under the stress-strain curve

Figure 1.

Cost" assembly method. A small supply of larger repair fasteners can be supplied to replace those in the few products that area drilled out or otherwise removed for repair purposes.

If ease or speed of disassembly or fastener reusability is important, threaded fasteners are often the answer. But there are many others, and the true fastener professional helps the customer make a thorough evaluation. Headed parts, pins, quick release fasteners are often candidates, to name a few.

The application often dictates the type of fastener needed. For instance, headed screws and bolts generally perform best when carrying loads in tension. Threaded products without heads, like set screws, are designed to apply compressive loads. Pins, shoulder screws and others with generally smooth cylindrical surfaces are designed to carry shear loads.

Obviously, the best design efficiency is obtained by holding parts together with the type fasteners that best support the load. For example, we would normally avoid placing a shear load on a tensile fastening. This is demonstrated in Figure 2.

Choose Grade and Size

Once the fastener type has been selected, size and placement can be decided. Efficient design calls for a minimum of fasteners to hold the assembly. Use of fewer screws results in fewer drilled and tapped holes. Also, with the addition of stiffening ribs, or a slight increase in flange thickness, the assembly can be made more rigid, thus reducing fastener tensile and bending loads, and reducing the size or number of fasteners required.

An investigation of design efficiency should include a check on the grade of fastener (see Figure 3). On the surface, it may appear as though it is least expensive to purchase Grade 2 strength level cap screws. But often the lowest cost fastener per pound of holding power is one of higher strength.

Because of this, either fewer of the higher grade fasteners can be used, or smaller size screws can be specified. In either case, the cost of drilling and tapping is reduced and the assembly itself can be kept to an efficient size. Total assembly cost is reduced.

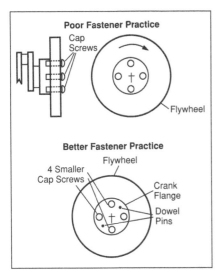

Poor Fastener Practice

Cap Screws

Flywheel

Better Fastener Practice

Flywheel

4 Smaller Cap Screws

Crank Flange

Dowel Pins

Figure 2.

This is an example of redesign for better fastener use. Originally a flywheel was mounted to a crankshaft using four cap screws, placing tension fasteners under tensile and shear loads. In the redesign, dowel pins oppose the shear load and the size of the cap screws is reduced because they are opposing stresses which are closer to purely tensile.

Take, for example, a situation where 60,000 pounds is the design clamp load. This would require (4) 7/8 diameter Grade 2 fasteners or (15) 3/8 diameter Grade 2 fasteners or (4) 3/8 diameter fasteners at 180,000 PSI tensile strength. Obviously, the product would have to be larger to accept the 7/8 diameter tapped holes; larger drill and tap cost is approximately 7 times as much; and power required to tap the hole is significantly higher. The higher strength screws provide the best cost assembly.

In some cases, joint materials may not be capable of withstanding loads applied by high strength fasteners, and in such cases only a moderate increase in strength grade may be possible without an unduly large increase in bearing area. There may be cases, such as in a flange fastening situation, where a large number of fasteners is desirable because they spread the clamp load over a large area. Nonetheless, use of higher grade fasteners may be cost effective in many situations.

Critical factors in efficient assembly and long-term operation include correct type and size of fasteners, proper location (to ensure equal load distribution) and proper installation. In applications which experience high cyclic loading, analysis of fastener loading is critical. Based on the relative stiffness of the assembly and the screw, and the external load applied, the fastener size and proper preload can be determined.

Preloading

To effectively hold an assembly together, tensile type fasteners need to be elongated. In most rigid joint/cyclical load applications, the screw material should be strained so that induced stress (preload) is greater than the service load.

Preload is normally achieved by tightening the screw. The torque required to achieve the specific preload depends on surface lubricity at heads and threads, surface finish, thread pitch, and head bearing configuration. Torque wrench reliability and repeatability, commonly known as gage R & R in SPC language, is also a factor.

When a fastener is tightened, the load on the assembly and fastener increases. In the ideal case, the screw stretches and the joint compresses. All this should normally take place within the elastic limit.

Suppose a joint has been tightened to a preload P_1 and additional load, P_2, tending to separate the members is applied. In general in rigid assemblies, as long as the external load is less than P_1, it has little effect on the tension in the screw. Even if such a load is frequently applied, the fastener generally will not fail from fatigue.

If, however, an external load greater than P_1 must be applied, it should be kept to a minimum, since it produces cyclic tensioning in the screw and may lead to fatigue failure.

This principal has important practical applications. Often when fasteners fail, the fix first tried is to switch to the next larger diameter. This involves changing the tooling for the hole preparation and tightening, and possibly changing assembly fixtures. Most likely, tightening the fastener to increase the preload would be a more logical, and economic approach. This represents the Best Cost Assembly Approach.

As a reference, a 180,000 PSI tensile strength fastener may have a mean endurance limit of only 10,000 PSI. This means that the fastener is capable of a maximum one-time stress of 180,000 PSI, but that a stress change felt by the fastener of more than 10,000 PSI could

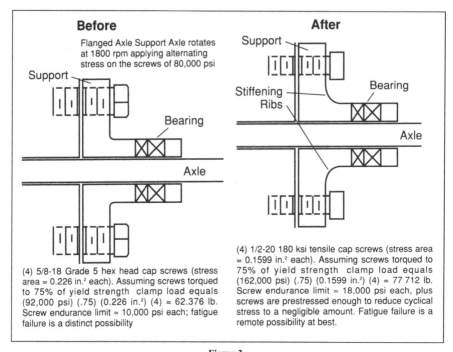

Before

Flanged Axle Support Axle rotates at 1800 rpm applying alternating stress on the screws of 80,000 psi

Support

Bearing

Axle

(4) 5/8-18 Grade 5 hex head cap screws (stress area = 0.226 in.² each). Assuming screws torqued to 75% of yield strength clamp load equals (92,000 psi) (.75) (0.226 in.²) (4) = 62.376 lb. Screw endurance limit ≈ 10,000 psi each; fatigue failure is a distinct possibility

After

Support

Stiffening Ribs

Bearing

Axle

(4) 1/2-20 180 ksi tensile cap screws (stress area = 0.1599 in.² each). Assuming screws torqued to 75% of yield strength clamp load equals (162,000 psi) (.75) (0.1599 in.²) (4) = 77 712 lb. Screw endurance limit ≈ 18,000 psi each, plus screws are prestressed enough to reduce cyclical stress to a negligible amount. Fatigue failure is a remote possibility at best.

Figure 3.
This is another example of redesign to make more efficient use of fasteners. A flanged axle support was changed to include stiffening ribs, and the fasteners were changed from 5/8-18 Grade 5 hex head cap screws to 1/2-20 180 ksi tensile cap screws.

result in a fatigue failure within a given number of operating cycles of the product.

A conservative formula giving tension on a fastener which has an external load P_2 is:
$Pt = (K_S/K_S - K_C) P_2 + P_1$

Where:

Pt = Total bolt load, lb
P_1 = Preload, lb
P_2 = External load, lb
K_S = Screw spring constant, lb/in
K_C = Assembly spring constant, lb/in

The spring constant, K, for an assembly member is given by:

(area, in²) (Modulus of elasticity, PSI)/(length, in)

Since these calculations ignore factors such as bending, heat, impact loading, or high frequency secondary forces, they present the fastener professional with an opportunity to shine, if they have the brilliance of product knowledge in these areas. Applications where the forces, pressures, temperatures, frequencies or atmospheres create unique and challenging problems for the fastener required to hold it all together. The distributors who have the technical tools to respond to these special requirements can reap the rewards of successful and efficient fastener applications. ❏

Reprinted from *American Fastener Journal,* July/August 1992

Locking Out Vibration

By Jim Speck, P.E.

One of my favorite examples of vibration induced loosening of fasteners was a pair of carriage bolts and wing nuts on my first lawn mower. It had a long tubular set of handles which acted like a tuning fork, a 3.5 horsepower engine, very sloppy thread fit with the fasteners installed normal (perpendicular) to the vibrational loading. With the low preload that could be developed with the wing nuts, vibrational loosening was practically guaranteed at the factory!

The solution involved a larger, wrench induced preload and staking of the tightened wing nuts. Not very pretty, but economical and effective—traits I required at the time, and still prize today.

In many vibrational loosening applications, you will find that a systematic, knowledge-based approach will serve you and your customers well. First, visualize the application as a system containing a vibration generator, tuning forks, fasteners and any dampers (or shock absorbers) such as gaskets, non-metallic parts and the like.

As the fastener supplier, your concern, and rightfully so, will be with the assembly hardware. But focus initially on the source of the vibration—the vibration generator. Most rotating or reciprocating machines have provisions for balancing by adding or removing mass, or by counterbalancing. Observations and suggestions along these lines can't help but to put you in good light with the customer.

Next consider the relative arrangement of the assembly members and their sizes. Vibrations have a wave type of propagation and transmission. Fastener sites which coincide with period locations of maximum vibration amplitudes will require the fasteners to resist additional loosening forces. A slight shift of fastener location can reduce this effect. Likewise, orientation of the fastener primary axis and wave transmission axis should be arranged to give maximum attenuation of the vibrational energy. Normally, a 45 degree inclination works best with threaded fasteners. Normal (perpendicular) or in line (parallel) alignment allows the wave like vibrational energy to be transmitted along the helix of the thread much more easily, causing unloading of the preload and possible loosening. Dampening in the form of nonmetallic, elastic-like members in the assembly have an advantage in this regard, even though they often present preload retention problems in more rigid applications. The customer's engineering department can be helpful in this area if approached properly, by giving you valuable information about any past history with the application and their own preferences.

Don't automatically assume that a traditional head and thread fastener—with a locking element—is the optimal solution.

And finally, carefully consider your fastener proposal(s). In a vibrational loosening application, the fasteners are working a double shift. In addition to applying clamp load, they have to resist loosening.

Don't automatically assume that a traditional head and thread fastener—with a locking element—is the optimal solution. Pins, both straight and tapered, as well as screws with shoulders or special anti-rotation features are viable alternatives. Use your creativity—and the knowledge you possess—to expand your customer's range of options. And no matter what the alternatives, do three things if at all possible:

1) Together with the customer, define success. What vibrational test results are satisfactory—be it 1st on, 5th off torques, transverse vibration test seconds, or whatever.

2) Test all the alternatives in a professional manner and without bias. Use statistics to analyze your data.

3) Present your recommendation based on the alternative or alternatives which have the best combined scores of vibrational results and purchasing economics.

That's the Best Cost Approach to Assembly Efficiency. ❑

SKILLKEEPER

From a tensile test ultimate fracture load, calculate ultimate tensile strength (UTS).

Data:

#8-32 Cap Screw

Ultimate Fracture Load = 2,500 lbs

Failure in the threads

Thread Stress Area = 0.014 in²

Solution:

2,500 lbs/0.014 in²

= 178,571.4 lbs/in² or 178,571.4 psi

Reprinted from *American Fastener Journal,* September/October 1992

The Universal Nature of Technology

By Jim Speck, P.E.

I've been running some of our headers at night. Several of us have been here at Crescent Manufacturing for the last few months. I suppose it is the ultimate extension of "hands on" management. We've been checking parts, changing punches, drawing wire—all the usual things. It gives you plenty of time to think. Two thoughts that occur to me are:
1) The industrial world is getting smaller.
2) Machines really don't care who operates them or what they produce.

Fax machines, express delivery and near instantaneous communication have become fixtures to such an extent that it is difficult to recall business of the past without them. We can fax back and forth to Brazil or Bangkok as readily as Bridgeport or Boston.

Components can be sent by UPS and a variety of other carriers with no unseemly difficulties. And while you can never replace the value of face-to-face meetings (or in your face as the case sometimes requires), we can, in fact, reach out and maintain many business contacts over long distance for a good length of time.

Which brings me to metrics, and machines. It seems that the recent trends toward a unified European market and the concurrent United States impetus toward widespread metric system usage through the Omnibus Trade legislation has generated both anticipation and anxiety.

From a manufacturing perspective, any anxiety toward increased metric usage is unwarranted. When a header, a roller or other machine tool converts raw material into a product or component, it has no thoughts or feelings on the subject. It does what it is set up to do, whether metric or inch.

And the same is globally true for all scientific and engineering efforts. Gravity exerts its force however expressed. Particles travel, accelerate and rotate regardless of the units. Technology is universal. Inch or metric—both are languages. They help describe our processes and projects along with the needs of our customers.

> **It seems that the recent trends toward a unified European market and the concurrent United States impetus toward widespread metric system usage through the Omnibus Trade legislation has generated both anticipation and anxiety.**

Just as an object in Europe, Asia or the Americas is the same, whether we describe it in French, Japanese or English, so dimensions, strength and the descriptions of forces are the same physical properties regardless of expression system.

And for headers, slotters and rollers, production is production. Whether in millimeters or inches. In fact, with many new measuring instruments, dimensions are given in either system with the flip of a switch. So if the machines don't care, and they don't—I've been spending a lot of time with them lately—those of us in the fastener profession should be at least as equally fluent in metric.

See below for some units which should be rehearsed until they become second nature. Being fluent in metrics is really just a matter of going the extra kilometer! ❑

FORCE

Newton   Pound

TORQUE

Newton-Meter   Pound-Foot

STRESS

Mega-Pascal   PSI (lb/in sq)

LINEAR DIMENSIONS

Millimeters   Inches

TOLERANCES

+/- 0.1mm   +/- 0.004
+/- 0.05   +/- 0.002

SKILLKEEPER

A customer requires minimum single shear strength on a 10mm metric dowel pin of 700 MPa. The test report from the mill reports the minimum force to shear a sample of the lot was 70,000 Newtons. The material is 10mm in diameter. Does the shear test data meet the specs?

Solution:
Yes. For a 10mm diameter, the 70,000 Newton single shear exceeded a shear stress of 700 MPa.

Reprinted from *American Fastener Journal,* January/February 1993

Fastener Friction

By Jim Speck, P.E.

Fastener friction is a lot like a tax. You know it is there, you probably can't eliminate it, so the best strategy is to minimize your losses to it. One of the best ways I know of doing this is to put numbers on the various friction components so that they can be measured and compared.

For this article, a grade 5 5/16-18 screw will be evaluated using five different friction characteristics. By observing the differences in tightening efficiency, we can see how assembly friction conditions influence clamping loads.

We start off with some fastener math. The torque-tension equation is fairly sophisticated and includes factors for bearing as well as thread components, and makes trigonometric corrections. Fortunately, we can use a simplified version, $T = K*D*P$ which will give us satisfactory accuracy for a majority of applications.

The equation specifies:

T for torque in
inch-pounds = K for friction (dimensionless)*
          D for nominal major diameter in inches)*
          P for clamp load (in pounds)

5/16-18 Thread Stress Area = 0.0524 in sq	
Grade 5 Strength Level	120KSI
Load	6288 lbs
Load @ 80%	5030 lbs

K evaluated at 0.5, 1.2, 1.9, 2.6, 3.3
using 0.7 increments

$T = K*D*P$

    $T = 0.5\ (.312\ \text{in})\ 2{,}000\ \text{lbs}$
       $= 312\ \text{in lb}\ (26\ \text{lb ft})$
    $T = 0.5\ (.312\ \text{in})\ 5{,}000\ \text{lbs}$
       $= 780\ \text{in lb}\ (65\ \text{lb ft})$

$T = 1.2\ (.312)\ 2000$	
$= 748.8\ \text{in lb}$	62.4 lb ft
$T = 1.2\ (.312)\ 5000$	
$= 1872\ \text{in lb}$	156 lb ft

$T = 1.9\ (.312)\ 2000$	
$= 1185.6\ \text{in lb}$	98.8 lb ft
$T = 1.9\ (.312)\ 5000$	
$= 2964\ \text{in lb}$	247 lb ft

$T = 2.6\ (.312)\ 2000$	
$= 1622.4\ \text{in lb}$	98.8 lb ft
$T = 2.6\ (.312)\ 5000$	
$= 4056\ \text{in lb}$	338 lb ft

$T = 3.3\ (.312)\ 2000$	
$= 2059.2\ \text{in lb}$	171.6 lb ft
$T = 3.3\ (.312)\ 5000$	
$= 5148\ \text{in lb}$	429 lb ft

While the equation and calculations suggest linearity, in actual assemblies some curvature to the function may take place due to deflections.

Figure 1.

For our 5 Torque-Tension Plots D will be 0.312 inches (5/16-18), while P and T will be plotted on the X and Y axis' respectively.

The tabulated calculations are shown in the Torque-Tension figure, starting at K = 0.5 and increasing in increments up to 3.3. Clamp loads of 2K and 5K pounds are used to set our torque-tension lines.

As can be seen in the torque-tension plot for K = 0.5, a small input of tightening torque gives a large gain in clamp load. This low a K value

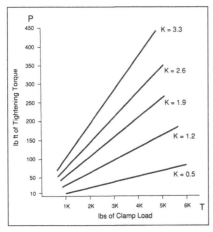

Torque-Tension

would be very efficient in tightening and would be associated with assemblies coated with very slippery materials such as molybdenum disulfide or similar "super" lubricants. It also would be efficient in promoting loosening if the assembly was subjected to vibrational forces.

Scrolling through the five plots in Figure 1 you can see that as K value increases, it takes more torque to develop clamp load. In practice the K value of 1.9 to 2.6 most likely bracket values found in actual practice. But grit, foreign material or any contamination which increases friction, and the K value, will cause a decrease in tightening efficiency.

By using torque-tension plots, we can give fastener specifiers and users a real view of what is taking place in their assemblies. And that can ne plotted using standard fastener lab equipment. Another example of the Best Cost Method of fastener engineering. ❑

SKILLKEEPER

An assembly line foreperson shows you an assembly oil used in assembling conveyor rails and asks for your clamp load estimate for the 1/2-13 Grade 8 cap screws they are using. The data sheet on the oil gives a K value of 0.2 and you are told that a torque of 1200 inch pounds is being applied.

Solution:

$T = K*D*P$, solve for P

$P = T/K*D$

$P = 1200$ in-lbs$/0.2(0.500$ in$)$

$P = 1200$ in-lbs$/0.100$ in

$P = 12,000$ lbs or 12K pounds

Reprinted from *American Fastener Journal,* March/April 1993

Fastener Fundamentals

By Jim Speck, P.E.

The fastening and assembly products market nationwide is worth over $7 billion, based on available economic statistics. Similar aggregate market estimates are available for the car and computer businesses; the electric motor indus- try; and just about every commercial segment you could contemplate. You can draw a mental image of cars, computers and electric motors. But how do fasteners appear to you?

Question 1.

How does a fastener work?

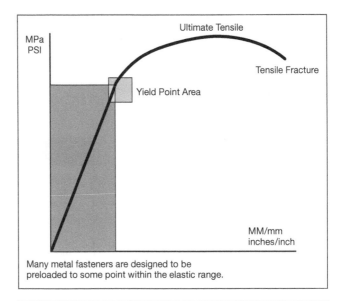

Many metal fasteners are designed to be preloaded to some point within the elastic range.

Question 2.

What are its mechanical limits?

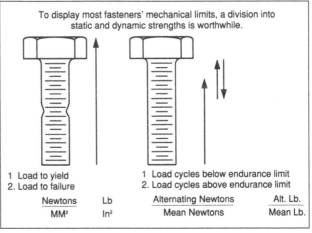

To display most fasteners' mechanical limits, a division into static and dynamic strengths is worthwhile.

1 Load to yield
2. Load to failure

1 Load cycles below endurance limit
2. Load cycles above endurance limit

Newtons	Lb	Alternating Newtons	Alt. Lb.
MM²	In²	Mean Newtons	Mean Lb.

Question 3.

Where is it operating most efficiently?

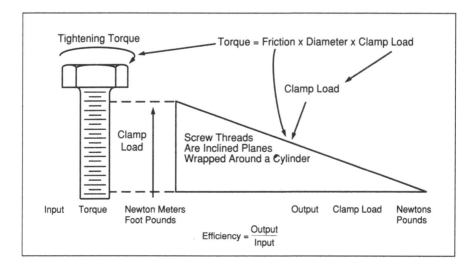

If you carry the general perception held in industry, this vision is of bolts, nuts, screws and rivets. Such a narrow view misses a profitable segment of our market. It also misses the needs of the most valuable observer, the customer. While you may be seeing—and selling—bolts, nuts, screws and rivets; they are seeing—and buying—holding ability. They're buying those products, processes and assemblies that hold their products together.

In the final analysis, this is what the fastener market is about—holding power and efficiency. In the long run, the parts which deliver the most holding power for the least cost, win the business. Three fundamental questions for those of us new to the business—or for the well-trained veterans—are:

- How does a fastener work?
- What are its mechanical limits?
- Where is it operating most efficiently?

Since a picture is worth a thousand words, let's use industrial pictures—graphs and diagrams—to explore each of these three questions. And since we operate in a global economy, let's include metric values in addition to inch system units, to gain familiarity. ❏

SKILLKEEPER

A customer asks how much torque to apply to get a 75% preload with M12, 12.9 strength socket head cap screws in an assembly of steel plates. Use a K value for friction of 0.2.

Strength level 12.9
proof strength 970 MPa
(from published data)

0.75 x 970 MPa = 725.5 MPa
= 725.500,000 Pa

M12 stress area = 84.3 mm²
= 0.000084 M2

Solution:
Target Preload in Newtons
725,500,000 Pa x 0.000084 M²
= 60,942 Newtons

Required Torque at K = 0.2
  T = K x D x P
  T = 0.2 x (0.012M) 60,942 N
  T = 146.26 Nm

146 Newton-Meters of Tightening Torque

Reprinted from *American Fastener Journal,* July/August 1993

Fastening With Pins

By Jim Speck, P.E.

Often we want to automatically think of screws or bolts when thinking about fastening. Industrial grade pins can be a more efficient choice for holding an assembly together. As with most fastening jobs, good pin selection starts with a thorough knowledge of the assembly and the service it is to perform. The discipline of Design for Assembly, as documented by Boothroyd and Dewhurst, gives anyone studying as assembly for possible pin application a powerful concept by asking several questions. They are:

- How do the parts *move?*
- What *materials* are used?
- When and where are they accessed for any *maintenance?*

One way to think of these qualifying questions is as the three M's: Move, Material and Maintenance.

Let's look at each, in turn, with respect to pin fastened assemblies.

Move

If the parts move—how? There are two primary types of movement. Translation, which is back and forth. And rotation, which is around and around. Some examples of each are:

Translation
Slider on a shaft
Machine slide in ways
Removable pin in/out of the site

Rotation
Collar around a shaft
Wheel around an axle
Hinge around a hinge pin

Obviously, if the movement is part of the design, the pin selection needs to take this into consideration. Less obviously, but equally important, intelligent design selection can provide years of successful application, use and sale of the product by your customer, and a steady stream of pin sales for you and your distributorship.

Materials

This second "M" is often given less consideration than it warrants. Chalk it up to human nature. It is easier to say steel, zinc and clear than to carefully think through the application, all possible candidate materials, weigh alternatives and make a considered decision. However, pin material and finish should offer the best finish, lubricity, strength, cost, availability and overall performance that can be summoned. Site materials, especially plastics, come in such a wide range that it can be a challenge to keep up with them.

But knowing the fastening site material's characteristics and what has worked well with it in the past can turn an average assembly into a great performing one—with improved performance by whatever measurement criteria is used combined with a lower installation cost. And the materials industry is rich in sources of information on materials. Try to cultivate materials sources as fountains of knowledge on the applications of their materials. I know of one pin application which went from good to great by just the change in nickel content of the pin raw material.

Maintenance

Surprisingly, in what we used to think of as a throw-away society, maintenance and recyclability are hot topics. Many of the car companies are pushing the development t of "green" products. (It's enough to make an engineer grin. We like engineering development like an accountant likes spreadsheets!) Many other industries will be picking up this trend. Products which can be recycled into fresh raw materials when they reach the end of their productive life will require simple, efficient fastening methods and pins are a natural in this regard. Disassembly is quick if designed properly. Which makes this third "M"—Maintenance—especially important if our pin fastened assembly needs to be efficiently disassembled. The capability to be removed has to be designed in to the pin and site. Strength, appearance and reusability are three of the prime factors in fastener selection. And reusability used to be one of the main reasons for the wide spread application of threaded fasteners such as nuts and bolts. Pins, however, can also be reusable. If the site and the site material are capable, a simple interference fit can be pushed back apart for servicing. In fact, if you study the average cold header or roll threader, it has many pin type fastened connections.

Other terminations for pins include spring type washers and press on retainers. Old hot rodder that I am, when I hear clutch I think small blocks and Hurst shifters, but simple spring clutches are also ideal for holding pins fast in an assembly and providing quick release for maintenance, service and recycling. Clevis pins, cotter pins, hairspring clips, roll pins... all provide the market with fasteners offering a wide spectrum of load carrying ability, appearance and installed cost. And speaking of load carrying ability, pins often are designed to carry shear loads. Shear loading is perpendicular (also called normal) to the lengthwise axis of the pin. Shear strength can be specified as either single shear or double shear. Other performance measures for pins are installation force, push out strength and surface finish.

Pins come in a wide range of shapes, sizes and strengths. They can accomplish an equally diverse domain of fastening and assembly tasks for your customers. ❏

SKILLKEEPER

A pin customer wants to use 300 series stainless steel for corrosion resistance. The single shear strength of the proposed pin material is 30,000 PSI (206 MPa). The pin diameter is 0.118 inches (3mm). What typical single shear loads would you expect on a test report?

Solution:

Shear area $0.7854(.118 \text{ inches})^2$
$= 0.0109 \text{ inches}^2$

Single shear estimate 30,000 lb/in^2
$(0.0109 \text{ in}^2 = 328.1 \text{ lbs}$

Reprinted from *American Fastener Journal,* July/August 1994

Temperature Effects on Fastener Materials

By Jim Speck, P.E.

Occasionally I'll get a call concerning fasteners for use in applications where service temperature is something other than room temperature. The most recent was for use in securing banks of F.E.T.s, or Field Effect Transistors to a machine base. While forced ventilation can control the amount of heat generated by these devices, the application still literally gets as hot as a furnace.

Fasteners and fastening systems designed and installed in applications where the service temperature is other than room temperature can present some unique problems if the temperature effects are not taken into consideration. Both low and high temperature extremes can decrease fastener performance. We will concern ourselves with some elevated temperature effects by considering first the engineering aspects of heat, then looking at how it changes tensile strength, comparing the differences in different fastener materials thermal expansion rates and conclude with a brief review of the phenomenon of creep.

Heat

Heat is energy. Heat transfers freely by conduction, convection and radiation from a hotter area to any less hot areas unless inhibited by insulation. Temperature by whatever scale chosen is a measure of the amount of heat energy. Although steel industrial fasteners may appear to be static pieces of metal, it is important to remember that a good deal of molecular activity is taking place inside the fastener. In an application where the temperature is dropping, such as bridge hardware on an increas-

ingly cold winter's day, the outflow of heat reduces the ductility of the fasteners and can have a direct impact on their ability to carry load. At elevated temperatures, heat energy flows into the fastener's metallurgy. This increase in internal energy makes the fasteners more ductile to a point. As the temperature rises above that, yield and tensile strength can decrease if the original tempering temperature of the fastener is exceeded. Figure 1 shows a graph of tensile strength and temperature for an alloy steel fastener. It is interesting to note the slight knee in the downward slope of the curve and that at elevated temperature tensile rate decreases quickly. As an example, alloy steel fasteners with a nominal hardness of HRc 40 and tensile strength of 180 KSI, would have an original tempering temperature of around 760 degrees Fahrenheit. If these fasteners are installed in an application which reaches an operating temperature of 900 degrees F, an additional 150 degrees F of tempering of the fastener can occur, with a corresponding decrease in strength.

Of equal concern with elevated service temperature is the duration of the elevated temperature. In addition to tempering or "drawing back" of tensile strength, two other factors of importance are the expansion rates of the assembly materials and creep, or gradual relaxation of load with time at temperature.

Heat transfers freely by conduction, convection and radiation from a hotter area to any less hot areas unless inhibited by insulation.

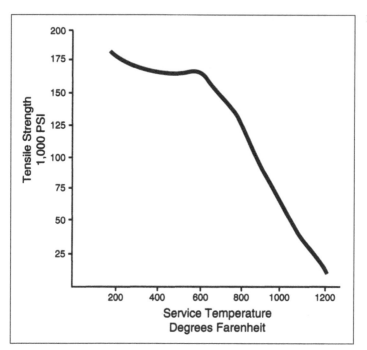

Figure 1.

Differences in Linear Expansion

With changes in temperature (ΔT), different materials contract and expand at different rates (ΔL). In elevated temperature applications, both the fastener and assembly expand. Each expands at the expansion rate of their material. This is usually given by the material's linear expansion coefficient. Table 1 shows the coefficient of linear expansion for aluminum, brass, cast iron, and carbon steel. As these "per unit" numbers are quite small, linear expansion rates for a temperature rise, or ΔT of 300 degrees Fahrenheit are also shown for these four materials.

Returning to our F.E.T. application, if a 300 degree F temperature rise occurs, both the fas-

tener and fastened materials will expand according to the differential expansion rates. In Figure 2 a steel screw holds an aluminum member to a cast iron base. Using an M3 sized machine screw with slightly over 6mm of grip length and 340 MPa of preload, the initial preload elongation is shown. By comparing the different expansion rates with the preload, it can be visualized that the potential for significant changes in clamp integrity of this assembly exists at the higher service temperature. As service temperatures rise, heat resistant fastener materials may be called for. Two popular heat resistant materials which are also cold headable are A-286 and H-11.

Linear Thermal Expansion

Material	Linear Expansion per Degree F	Temp.	Linear Expansion
Aluminum	.00001244 inch	300 F	0.0037 inch
Brass	.00001 "	"	0.0030 "
Carbon Steel	.00000633 "	"	0.0019 "
Cast Iron	.00000655 "	"	0.0020 "

Table 1.

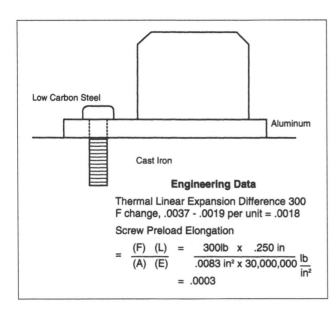

Figure 2.

Low Carbon Steel

Aluminum

Cast Iron

Engineering Data

Thermal Linear Expansion Difference 300
F change, .0037 - .0019 per unit = .0018

Screw Preload Elongation

$$= \frac{(F) \ (L)}{(A) \ (E)} = \frac{300lb \ \times \ .250 \ in}{.0083 \ in^2 \times 30,000,000 \ \frac{lb}{in^2}}$$

$$= .0003$$

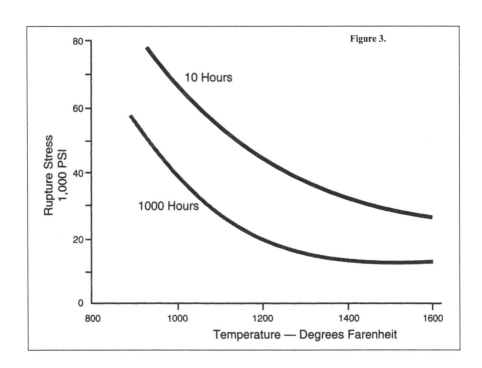

Figure 3.

What makes creep of particular interest is that the loads to cause plastic, or permanent, deformation of the fastener are lower than would normally cause assembly problems.

Creep

Of final concern is the loss of fastener load during sustained load at elevated temperatures. It is this gradual loss of load with time at temperature, known as creep, which can have an impact on elevated temperature service fastener applications if it is not accounted for during fastener specification. What makes creep of particular interest is that the loads to cause plastic, or permanent, deformation of the fastener are lower than would normally cause assembly problems. It is the combination of elevated temperature and loading which together cause the fastener's material to creep under load. A plot of stress versus time to rupture similar to the one shown in the figure can be obtained either from engineering sources for published data or from laboratories equipped for this testing. Good information on the load effects of fastener and fastened materials at service temperatures and beyond is of specific interest if this type of service is a possibility.

When it comes to fastener materials for these "Hot Jobs," attention should be given to the additional work the fastener must be designed to perform. ❑

Reprinted from *American Fastener Journal,* May/June 1995

Thread Pitch's Influence on Assembly Speed

By Jim Speck, P.E.

One of the advantages of using threaded fasteners is the degree of reusability they offer. They can be tightened and untightened for as many cycles as an application's service and in the end, recycling, require. Especially in these times when the recycling of both plastic and metal components is both desirable from an economics viewpoint and helpful in conservation efforts to make the most efficient use of raw materials.

A good way to think about screw threads, whether they are machine screw threads used in metal or spaced self-threading screws forming into a plastic site, is as an inclined plane wrapped around a cylinder. Seen in this manner, their mechanical advantage is easier to discern. If you push a wedge under a door, it pushes up on the bottom of the door, holding the door open. On a tension fastener such as a cap screw, with right-handed threads, when the screw is turned clockwise, it moves into the assembly until it starts to clamp the parts together. Continued turning results in more advance, but now with more work being done as fastener and parts are being stretched and compressed. This of course results in more torque being required to continue with the turning of the screw. It is worthwhile to understand the connection between degrees of threaded fastener rotation and the resulting lead, or advance. Figure 1 can be used as a model to show the relationship of turning to advance. And of course in the loosening direction, either through untightening or vibrational loosening, the direction of travel is reversed.

As can be seen, the thread "helix" angle, the thread diameter and thread pitch all work together. I think the metric system of calling out the true thread pitch shows this much better. For example, an M6 x 1 advances 1mm for every 360 degrees of rotation. For a 1/4-20,

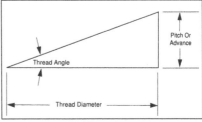

Figure 1.

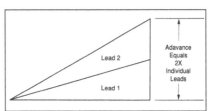

Figure 2.

you have to know that 20 TPI, or threads per inch, figures out to 1/20th of an inch per pitch or 0.050 inches for every 360 degrees of rotation. This is for threads with one thread start. If we make the threads two starts, a "twin lead" thread, for each rotation, we would advance 2x the lead. Again, for an M6 x 1 twin start thread the advance would be 2mm. Figure 2 shows the graphic presentation of this effect.

As can be seen, the two leads advance the fastener at double the rate of the same pitch, or thread per inch, as a single lead. I have also found that on small self-threading screws, two thread starts help stabilize and balance small diameter, longer length screws as they start into cored, injection molded plastic holes. This faster tightening rate, with fewer revolutions to seat, can be used to advantage in some applications. It can make for a nice gain in assembly efficiency. They do have some engineering drawbacks so it is important to understand the application well before specifying multiple start threads. Vibration resistance can be lower if vibrational energy service loads are part of the application. On machine screw threads, especially finer pitches, they are easier to cross thread with all the problems that can bring. Roll thread die life can also be lower. But used in the right applications they can give a real boost to assembly output. One triple lead electrical connector screw I am familiar with closes a 0.375 inch grip length gap in a turn and a half. The connector literally flies together when you tighten the screws and it still develops hundreds of pounds of clamping force! ❑

SKILLKEEPER

A special 1/4 inch diameter screw has 18 threads per inch. What is the lead advance per 360 degree rotation?

Solution:
18 divided into 1 inch = 0.0556 inches per 360 degrees of rotation

Reprinted from *American Fastener Journal,* July/August 1998

A Socket Screw Application Remembered

by Jim Speck, P.E.

As this issue of *AFJ* features socket screws and is a socket products reference issue, and Mike McGuire reminds me that ten years ago when the socket screw reference issue was last visited, I thought that the following true story might be of interest.

A manufacturer of construction equipment designed, manufactured and was marketing a line of flat bed truck mounted cranes. These were used by forest products companies to pick felled and pruned tree trunks for transportation to saw mills. It was a well received product which had been giving a good account of itself in the rough and tumble surface encountered in the back woods. Figure 1 provides a sketch of the crane in side view.

Well over one hundred of these units had been purchased and were operating in the field when failures started to occur in the 7/8 inch diameter socket head cap screws which held the rotating crane boom to its heavy steel base. These were long screws, and the failures were right at the head/shank fillet. By the time we were made aware of the application problems by our distributor, several of these crane booms had fallen off under load in the field. Fortunately, no one had been injured.

In looking at possible failure modes, it appeared that some of the socket screws had fractured right at the head-to-shank fillet. A metallurgist evaluated the mechanical properties of the parts. What was found was fatigue failure. The gross chemical analysis of the fractured parts showed correct analysis. The microstructure demonstrated proper heat treatment. More importantly, the fracture planes showed the characteristic "beach marks" which are typical of tensile fatigue failure. As a side note, it is important in working with fractured parts to avoid the temptation of fitting the fractured parts back together and rubbing the fracture planes against each other. This can burnish some of the tell-tail marks from the fracture and make a metallurgical opinion on the cause of fracture more difficult to be determined.

With fatigue failure of the socket screws being identified in the fastener lab as the probable cause, we next went to the crane manufacturing floor to see what we could observe. I believe, once equipped with enough application facts, that the process of closely observing the how, what, when, where and why of any application is a critical step to fastener problem solving. Informed observation combines lab hypothesis with application facts to equal the true effectiveness and efficiency of a fastening.

At the site, we were met by a capable, concerned but enthusiastic group determined to

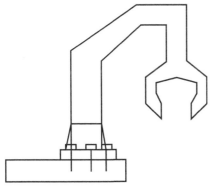

FIGURE 1.

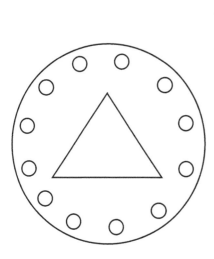

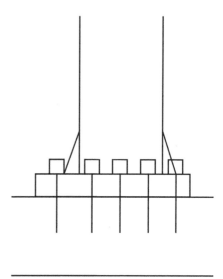

FIGURE 2.

find the cause of fatigue failure and implementing the most effective "fix." The socket screws were configured in a circular pattern as shown in Figure 2.

In Figure 2, I would like to draw your attention to two elements. First, note that the boom is triangular in shape while the bolt pattern is circular. Second, observe that the "anchors" at each end of the fasteners—the heads and the threads—were applying the fastening forces directly onto the bearing surface on the underside of the head and into the tapped holes in the crane bed frame. A team of engineers and technicians was brought together. It was decided that we needed to know the actual stresses on the individual screws. We were fortunate to have the use of an electronic load cell. We wired the individual fastener locations, one by one, operated the crane over a full range of operations and recorded the data. Two things became much more clear. The screws were not carrying the load equally due to the triangular nature of the boom. The three fasteners at one apex of the boom's geometry showed peak loads considerably larger than the rest of the screws in the bolting pattern. And those peak loads, while below the yield strength of the screws, were in excess of the tension-tension fatigue strength.

There are usually several methods to solving a fastening problem, any one of which will return fastening integrity. The key is to select the approach that is most efficient in time and materials.

Our log picking crane had a happy ending. It was obvious that the three screws located at the critical boom apex required more strength than the existing fasteners. Also the tapped threads were not capable of carrying higher fastening loads nor was the bearing area on the under head area of the socket screws being adequately carried by the mating boom flange surface.

Those three sites had the tapped threads drilled out and longer socket screws with hardened tool maker washers and high strength locking nuts were installed and carefully preloaded. The strain gages were installed and full range of motion crane operation now showed peak loads well below the new fastener's endurance limit. The test unit and strain gages were left in place for a week's worth of running to prove out the fastening design change.

We were pleased to hear that crews dispatched throughout the United States and Canada were able to make field upgrades to all existing units while new production was revised for the improved hardware. ∎

Reprinted from *American Fastener Journal,* January/February 1999

When Should Fasteners Be Replaced?

by Jim Speck, P.E.

Is there a practical method for determining when it is advisable to use new fasteners in a machine that has been in service for some length of time? A key factor is the cost of an assembly failure compared with the cost of installing fresh fasteners. Let's consider some of the implications of this and then take a look at a hypothetical application. The cost of a failure can include direct, indirect and peripheral costs. They can include such examples as damage to the assembly itself and costs to get back into operation. Examples would be broken assembly members related to a fastener letting go. Indirect costs would be those arising from costs to get back into operation. For example, if a technician needs to be brought in on overtime to make the repairs made necessary by the fastener failure, those costs would be "costs of failure" of the indirect variety. Finally, peripheral costs would be those attributed to the failure in ways that may not be directly or indirectly obvious. Obvious would be if someone could be hurt physically or otherwise by the fastener failure. Less obviously, let's say word of the failure travels the rumor mill circuit. This can either be inner-company or in the customer market. A side note is the e-mail and internet access are rapidly compressing the "time to market" for any newsworthy bit of information. And often the best communication policy for negative PR is not to have the news launch in the first place! In summary, a fastener's cost of failure can be thought of as comprising the direct, indirect and peripheral costs. And sometimes, the best thing an irate customer can do is complain to you. If you take the attitude that " I'm not hearing any complaints," it could be they are upset enough

that they have just stopped talking, and buying, from your company. The cost of failure can be expressed in equation form:

> **COST OF FAILURE =**
> Direct + Indirect + Peripheral Costs

On the other side of the reuse/replace decision, the cost of installing fresh fasteners includes more than just the purchase price of new fasteners. It also includes the costs to run them in and torque them up. This installation cost can typically be four times the procurement cost. Since these costs do not usually get typed up and delivered on an invoice, they are not as readily recognized. A popular way of expressing procurement costs are by the 18-82 Rule. 18% of the cost is to buy the fasteners and 82% of the costs are to install them. You might note that this is very similar to the ratio of the Pareto Principal which states that 80% of a project's cost will come from 20% of its items. If we express the upgrading fasteners cost it would as follows:

> **UPGRADING FASTENER COSTS =**
> Fastener Purchase Cost +
> Installation Cost

To continue our math model, if we find the point at which these two equations are equal, namely:

> **UPGRADING FASTENER COSTS =**
> COST OF FAILURE

We find that direct and indirect costs fairly well cancel each other out on both sides of the equation and we are left with a situation where the peripheral costs equal zero. In other words, if the peripheral costs of a fas-

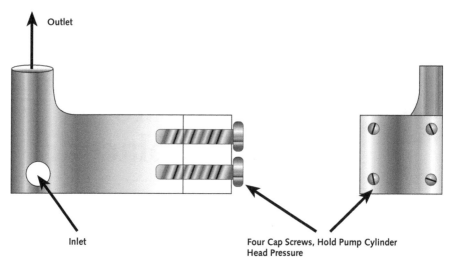

FIGURE 1.

tener failure after being replaced are nil, then reusing the fasteners can save some cost and might make good sense.

However, if the peripheral costs are an increment above zero, then replacing the fasteners rather than reusing them is probably the most reliable decision path, provided of course that the new fasteners are of equal quality and installed properly. Let's consider a hypothetical application. Figure 1 shows the sketch of a process pump used in a manufacturing process. After ten years of reliable use, the equipment, including the pump is to be rebuilt in place to upgrade the controls and renew worn bearings and seals.

The pump is fastened with four 300 series stainless hex head cap screws which cost about $1 each. Labor to the fasteners is around $18. If the pump fails, damage to the process equipment can run several hundred dollars. During the ten years, the screws have withstood around 7.5 million loading cycles. You are in charge of the operation—You make the call!

Using the math models from above, what it points to is, you can save a few ten-dollar bills by reusing the existing fasteners, but add a risk of a several hundred-dollar loss. Seen in these terms, new fasteners would seem like the more prudent choice. Can we hedge our risk by learning how much life is left in the fasteners? Probably, but the cost to have them liquid penetrant inspected far outweighs the cost of fresh screws.

Why don't they last forever? Well, they might if the load cycles are varied in amplitude compared with the fatigue strength of the fasteners. And fracture mechanics can guide us to what discontinuities are troublesome and indicators of future failures.

What you come back to is a balancing of fastener replacement costs and "peripheral" costs of failure. And when a possible failure of used fasteners with an unknown percentage of their fatigue strength consumed is in your "peripheral" area, good practice may just be to install fresh fasteners! ∎

S K I L L K E E P E R

PROBLEM

A fastener user wants to determine how to plot the service life of a cylinder head screw. What information would be valuable to know?

SOLUTION

The endurance limit would be a useful statistic. It would specify limits such as the tension-tension endurance limit for a candidate fastener as an alternating load of 10 KSI for 10 million cycles.

Reprinted from *American Fastener Journal*, May/June 1999

Metrics and Miniaturization
THIS TREND IS YOUR FRIEND

by Jim Speck, P.E.

What happens when an assembly is downsized? For one, fewer raw materials are used to manufacture it. If it is portable, less work is needed to ship it, transport it and store it! Also, more features and consumer benefits can usually be packed into smaller envelopes of dimensions. So, as products are made smaller and lighter, they ironically also become more valuable and efficient. This seeming paradox is at the root of the rapid rise in the use of miniature sized fasteners. Add in the fact that many new products are designed to be sold around the world making metric dimensions a mandatory requirement, and you have a confluence of smaller screws, pins, etc. in mm units.

Three markers can be guideposts in this metric-miniaturization trend. They are:

1. THE RULE OF EXPONENTIATION

2. THREE FUNDAMENTALS OF MINIATURE FASTENER USAGE

3. THE SMALL SPACE PARADOX

Taken in order, the rule of exponentiation refers to the formula for circular area which raises diameter to the second power, or in other words squares it, to find the working area of either the roll thread and pitch diameter of threaded fasteners or the shank area of pin type fasteners. It can be simplified to:

Miniature metric fasteners need to satisfy three fundamental characteristics: strength, reusability and appearance.

$$\text{Fastener Area} = 0.7854(\text{diameter})^2$$

Let's take a change in fastener size. We are using a 4mm screw and our new specifications call for a 2mm diameter. Since the diameter is halved, what happens to our circular area? If you guessed halved—you lose! The new area is one fourth! Let's do the math:

First Area = 0.7854(4mm)^2 = 12.57mm^2

Second Area = 0.7854(2mm)^2 = 3.1416mm^2

3.1416/12.57 = 0.25 or one fourth

The implications of this are that as fasteners get smaller, the strength of their materials and precision of their dimensions goes up exponentially. The more generous margins of larger diameter parts are just not there in miniature parts.

Miniature metric fasteners need to satisfy three fundamental characteristics regardless of their feature size, namely: STRENGTH, REUSABILITY, APPEARANCE.

Let's consider an M1.6 machine screw threaded into a threaded insert in the injection molded chassis of a small, lightweight digital telephone. The screw and insert need to be capable of generating sufficient clamp load to keep the components secure through temperature changes, occasional rough handling and possible vibration and plastic creep over time. Should service or recycling be required, the unit should allow for ready disassembly. And the fastened assembly needs to achieve a quality, well put together presentation so that the sometimes fickle and often fleeting attention span of every potential customer can be arrested and drawn to our product and away from the competition's product.

Finally, let's consider the paradoxical demands a very small assembly site places on the fastener and the distributor who supplies them. The drive, be it cross recess, internal wrenching, slotted or what have you, may be much more difficult to reach with a wrench or driver. By necessity they must be not only to industry specification, but the best that can be supplied. Since the area being clamped is very small, the fastener bearing area must be as fully formed and perpendicular to the thread axis as fastener making permits. And since the mating hole will be tiny, the thread lead in and starting thread(s) must be well defined so that no cross threading or off-angle starts occur.

To summarize, using small metric fasteners requires combination of attention to details and precision in design and usage. Given this, there are big fastener gains to be had from using small fasteners! ■

SKILLKEEPER

An application uses an M1.4, internal wrenching machine tapping screw with an average nut factor of 0.2 and a tightening torque specification of 130 Newton meters. On average, how much clamp load will be generated?

SOLUTION

460 Newtons

Reprinted from *Fastening*, December 1999

Socket Set Screws

Where One Good Turn Deserves Another

BY JIM SPECK, P.E.

Set screws are really neat!! They are compact, work with reverse forces (they are compression fasteners) and can sometimes be installed so stealthily that you do not even know they are there. Unless of course they stop working—can you say loose collar? Yet, for all their ubiquitous nature, not a lot is written about them as their headed and threaded cousins. Let's take a look at how socket set screws work and explore some interesting features about this low profile but potent fastener type. Figure 1 shows a classic cup point socket set screw about two thread diameters in length, as well as a set of tightening torque and thrust forces as the input and outputs.

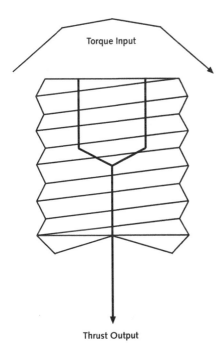

Torque Input

Thrust Output

FIGURE 1.

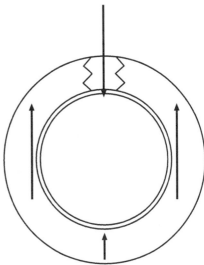

FIGURE 2.

A good example of a typical two-diameter length would be a #4-40 x 1/4 socket set screw. As can be seen in the figure, as tightening torque is applied to the socket, thrust is developed at the point of the screw. Kind of like a jet engine. This thrust can be put to good use in holding a collar tight on a shaft. Let's take a look at this set screw in an application. We will say the set screw is installed in a pillow block that fastens a power transmission shaft in a piece of packaging machinery. We have applied the full amount of tightening torque the hex key and socket are capable of developing. The resulting compressive force through the point and onto the shaft is the compression equivalent of the elastic stretch of a tension fastener with a head and threads. Let's look at the forces and reaction in both the screw and the collar. These can be seen in Figure 2.

Notice in Figure 2 that the reaction of the pillow block collar is to stretch into an elliptical shape. This stretching reaction is the action that generates the holding power that helps fasten the collar to the rotating shaft. Think of it as an oversized metal elastic band. The set screw stretches it and it pushes against the shaft in trying to return to its original round shape. Making a model of the forces, we would have a sketch similar to the one in Figure 3.

Note that the set screw and collar work as a set to accomplish the fastening. If it is a matched set, the screw's thrust will balance well with the area and material strength of the collar wall. Two points of interest that are often potential problem areas in set screw applications are:

- The relative hardnesses and deformations of the screw point and the mating shaft surface.

- The increase in holding strength from using two set screws.

Let's consider these areas in order.

At the shaft surface where the set screw point makes contact, usually one is going to be harder than the other. By specification, socket set screws are HRc 48-52. If the shaft is cold rolled steel, the hardness is going to most likely be in the mid to high Rockwell B range and the set screw point will penetrate

FIGURE 4.

until an adequate footing is established. If, on the other hand, the shaft is hardened and ground, the shaft surface hardness is going to be in the high HRc 50 to low 60 range. In this case the set screw point is going to be "softer," although not by many points. A spot face or other suitable indent in the shaft, along with a corresponding geometry point will enable the compressive thrust to be developed properly.

With two set screws, what is the gain in holding strength? The tendency is to say double—but not so fast—it depends a lot on the angle between the set screws. If they are in line, then a max. increase may be seen, but if they are at an angle, the resulting thrust is the resultant of the individual screw's thrust vectors. Figure 4 shows a sketch of the twin set screw effect. ⚫

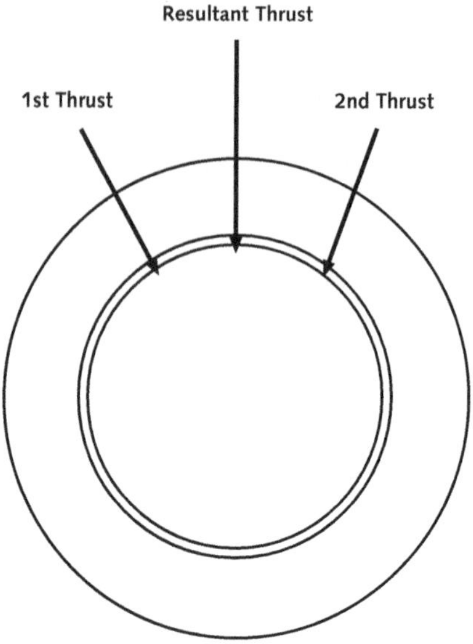

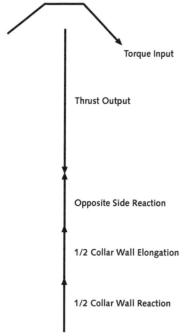

Torque Input

Thrust Output

Opposite Side Reaction

1/2 Collar Wall Elongation

1/2 Collar Wall Reaction

FIGURE 3.

SKILLKEEPER

A two set screw collar has an angle of 45 degrees between the set screw centerlines. Each screw develops 1,000 pounds of compressive thrust. What is the resultant thrust?

SOLUTION

1,847.8 pounds

Reprinted from *American Fastener Journal*, March/April 2000

Fasteners and Beyond

Where are we now and which way are we headed?

by Jim Speck, P.E.

Where does time go? It seems like just a few years ago that we were looking at Japan and wondering how they were pulling off the manufacturing miracles we kept reading about. Then SPC was going to change the way we reduced variability in our processes and lead to a better way of operating. Personal computers, software and the Internet have again promised to change our perspective on how we do business. But on a more fundamental basis, how have all of these "paradigm shifts" impacted the business of manufacturing, selling and using fasteners? People still use fasteners for three very basic reasons, namely: Strength, Reusability and Appearance. Let's make a three-by-three matrix of these factors and consider them in turn. The matrix can be seen in Figure 1.

How have the fastener factors been affected by these fundamental changes in business? Let's take strength first. Clearly, foreign competition in the form of fastener imports to North American fastener markets caused a dramatic disruption of the

	Foreign Competition	SPC	E-Business
Strength			
Reusability			
Appearance			

FIGURE 1.

markets. Some U.S. fastener manufacturers went out of business. Others merged. Almost all were forced to become more competitive in the use of higher speed equipment, better tooling and quick change set-ups as well as a realization that operator skill is a real factor in fastener quality, and useable strength. Attention was required to make sure that grade 8 strength really meant grade 8 strength. In a way, foreign competition caused two changes in fastener strength. One was the focus on fastener strength as a mechanical value that counted as one of the deliverables in a shipment of fasteners. The other was in the strength of the organizations that were delivering the goods. You had to be good or be gone!

How about SPC and the fastener strength? For a start, end users, distributors and manufacturers could all be on the same page when discussing fastener strength. In a report on fastener strength, a more universal grasp of statistics meant that not only the tensile strength readings could be reported, but also the mean, range and standard deviation of the test results. Process capability ratios of different processes could be compared to determine which were most efficient.

Finally, e-business tools such as e-mail and the Internet, not to mention such prosaic vehicles like the enhanced fax machines and targeted publishing, put lots of information in front of a lot of eyes—around the world. To sum up strength, market share was won by those suppliers who could deliver documented strength, with minimum variability anywhere in the world the assembly was being performed.

Let's take a look at how reusability fared. There was a renewed understanding that not all products, especially consumer electronics with relatively quick obsolescence cycles, need to be especially quick to be disassembled. The growth of single use fasteners and assembly methods grew dramatically. With assembly operations situated around the world, sometimes in areas of low assembly skills, fasteners that were threaded and meant for reuse required an ease of use to a much higher degree. Tooling and assembly systems also gained a greater attention. Here again, statistical methods meant that operations doing the same fastening work could easily be compared. And using modern electronic communications made survival of the fittest the law of the fastener jungle both for assembly operations and the fasteners and tools being used.

Finally, appearance of the final, assembled product took on a more critical role as consumers had the best of both worlds—a global range of suppliers. The near instantaneous ability to evaluate different product alternatives, often using numerical data like sales numbers, meant time between failure statistics, and other quantitative information, enabled only the most popular (read "best looking"), most cost effective and most reliable products to survive and prosper after their initial launch.

Strength, Reusability and Appearance are still three prime reasons why $28 billion dollars worth of fasteners and tooling are sold worldwide as we work in years starting with two with a couple of zeros for a while. We will all need to learn to work effectively in an increasingly globally competitive market which is numerically literate and electronically connected. In the fastener business we will be using all of our strengths, but reconfigured to these new realities. For those willing to forge ahead, these will be very interesting times. ■

SKILLKEEPER

A company requires a pin type fastener with a single shear strength greater than 900 Newtons. Figure 2 shows the application. Which of the following pound ratings provides this level of single shear strength?

A) 50 pounds

B) 100 pounds

C) 150 pounds

D) 250 pounds

SOLUTION

D) 250 pounds

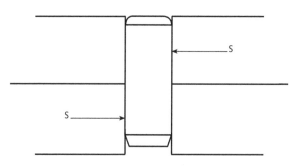

FIGURE 2.

Reprinted from *American Fastener Journal*, May/June 2000

Calculating Process Capability

Or "How Tight Did You Say You Could Hold That Tolerance?"

by Jim Speck, P.E.

Often we get asked what tolerance to apply to some features on special parts we are quoting. The conversation usually takes the form of "how does +/-0.005 inches (or +/-0.13mm if it is a metric part) sound to you?" "Yea, that sounds about right" is a typical acceptable response. This works better than one might think. But you know what? For those applications where it does not, being able to collect data and calculate some process statistics, including the Cpk, can save a lot of trouble. The educated guess approach is qualitative while the Cpk approach is quantitative.

Let's look at a relatively simple part, as shown in Figure 1.

To illustrate the use of process statistics to determine process capability, let's say that during a four hour production run of 50K parts, we measure the length of a group of parts every 15 minutes. If 50K parts in four hours seems high, it is because we are roll-forming these pins on a high speed rolling machine rather than producing them on a conventional screw machine. Since this will be a new manufacturing process for us, we do not have a wealth of past process data from which to draw. We will make each group contain five parts. Our total number of measurements equals 80. Table 1 shows the length data we have collected.

Notice that for each time we measured parts during this run, we collected data on more than one part. For this run we used five parts per sampling. We could have

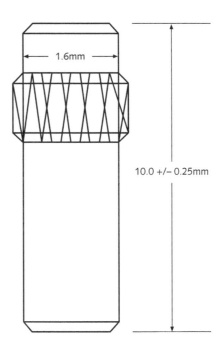

FIGURE 1.

designed our measurement plan for two, three or more. It is important for accuracy that we make a decision on a specific number and use it consistently throughout the run. We will select a factor we will use to calculate scatter in our data depending on this number. For the data we have collected we have figured the average of each group of five as well as the range for each group.

Why not just measure a few during the run to be sure we are staying within the specifications? In fact, we could and make perfectly satisfactory parts. The drawback would be that we would not have a clear understanding of the nature of our process' capability. It would also be less than precise in comparing length quality from one step-up and manufacturing run to the next. When we use the approach in Table 1 we are using a concept called the *Central Limit Theorem*. This statistical concept makes three fundamental points:

- Sample averages tend to be normally distributed regardless of the distribution of the individual measurements.
- The average of the distribution of sample averages will approach the average of the distribution of the individuals.
- The variance of the sample averages is less than the variance of the distribution of the individuals.

These points are predicated on one population of parts being sampled. Referring to the pin length measurements in Table 1 we can observe the average and the range for each group. Now let's calculate the average of the averages, which we call the Grand Average (or X Bar) and also the average range, which we will call the Grand Range (or R Bar).

Grand Average or R Double Bar =
Sum of 16 group averages divided by 16
= 10.16mm

Grand Range or R Bar =
Sum of ranges divided by 16
= 0.0669mm

Knowing these values, we can now calculate the measure of variance, or sigma and use that and the proposed tolerances to calculate the Process Cpk, which will be our measure of how capable our process is in holding these proposed tolerances.

Sigma = R Bar divided by 2.33
= 0.0287

Note we obtained the value 2.33 from a table of factors based on our sampling of five per time. If we had used four per or six per, our factors would be 2.06 and 2.53 respectively.

Similarly, we find our Cpk = minimum distance to one of the tolerances divided by 3 times Sigma:

Cpk = 10.25-10.16 ÷ 3(0.0287)
Cpk = 1.05

To say we have a capable process, Cpk should be greater than 1.5. In this case we

TABLE 1.

10.12	10.15	10.11	10.15	10.22	10.19	10.20	10.11	10.19	10.17	10.17	10.10	10.19	10.14	10.19	10.20	
10.17	10.11	10.19	10.21	10.13	10.20	10.17	10.19	10.15	10.17	10.16	10.14	10.15	10.14	10.11	10.17	
10.13	10.14	10.15	10.18	10.16	10.14	10.21	10.17	10.14	10.15	10.21	10.15	10.19	10.14	10.10	10.20	
10.20	10.16	10.13	10.17	10.15	10.13	10.13	10.14	10.14	10.13	10.18	10.19	10.14	10.16	10.16	10.14	
10.18	10.17	10.14	10.14	10.18	10.20	10.14	10.13	10.18	10.16	10.17	10.11	10.16	10.19	10.17	10.15	
50.8	50.43	50.72	50.85	50.84	50.86	50.84	50.83	50.72	50.80	50.89	50.76	50.74	50.82	50.68	50.85	**SUM**
10.16	10.15	10.14	10.17	10.17	10.17	10.17	10.17	10.14	10.16	10.18	10.15	10.15	10.16	10.14	10.17	**MEAN**
.08	.06	.08	.06	.09	.07	.08	.07	.07	.06	.04	.08	.05	.05	.07	.06	**RANGE**

```
                              X
                              X
                              X
                              X                 X
                              X                 X     X
                              X     X           X     X
                              X     X           X     X
                        X     X     X     X     X     X
                        X     X     X     X     X     X
            X           X     X     X     X     X     X     X
            X           X     X     X     X     X     X     X
            X           X     X     X     X     X     X     X
      X     X           X     X     X     X     X     X     X     X
      X     X     X     X     X     X     X     X     X     X     X     X

10.09 10.10 10.11 10.12 10.13 10.14 10.15 10.16 10.17 10.18 10.19 10.20 10.21 10.22 10.23
```

FIGURE 2.

are under this. To better understand what course of action to take, let's take a look at the distribution of individual readings, compared to tolerance. This can be seen in Figure 2.

Providing this distribution population is typical of the normal run for this process, if we recenter the process during setup to shift our averages to the mean length of 10.0mm, our Cpk will increase to a more acceptable level. Or, if our customer prefers that the parts be run on the long side, then retolerancing to 10mm, minus zero, plus 0.5mm will also have a desirable effect on our ability to consistently produce good parts. ∎

SKILLKEEPER

A small screw manufacturing run has an X Double Bar for head diameter of 0.119 inches, an R Bar of 0.001 inches and a sigma of 0.00043, using groups of five. The tolerance is .115 to .125 inches. Find our process Cpk.

SOLUTION

Cpk = 3.1

Reprinted from *American Fastener Journal*, May/June 2002

Run to Daylight

by Jim Speck, P.E.

Run to daylight. These three words tell a football running back what to do after the ball is snapped, the quarterback has made the hand-off and the offense is executing their blocking assignments. Noise and chaotic motion reign all along the line of scrimmage. Run to daylight guides the ball carrier to success no matter how the opponent counters. Run to daylight means look for success amid the chaos!

Technical subjects can be like a line of scrimmage after the snap. Lots of noise in the marketplace. Commercial projects and orders being performed with varying degrees of perfection. Knowing which way leads to success can make all the difference. Screw threads can cause some of the most perplexing quality issues in all of the fastener business. Are the threads "good" or "rejects?" The run to daylight rule for any fastener running back trying to make yardage in the fastener game is to look for the key blocks along which the route to success can be found. Let's start with the fundamentals.

Figure 1 shows external and internal screw threads in cross section such as the ones in your company.

Some of the terms we want to become understand completely are identified in Figure 1. On an external thread the Major Diameter is the straight line distance across the thread ridges. When threads are produced by roll threading, this is the last feature to be filled out. If there is a major diameter, there must be a minor diameter. As we can see, this is the linear distance across the thread grooves. Many of the terms used come from the gear industry,

which rapidly developed the engineering to perform precision measurement. From the gear industry we get the important term of pitch diameter. On a gear the pitch diameter has an addendum (distance above the pitch circle) and dedendum (distance below the pitch circle). This helps understand the pitch diameter of screw threads as well. The pitch diameter is the location where the width of the ridge equals the ridge of the groove. Any screw thread defined by a diameter and pitch, ½-13 or M12 x 1.75, for example, will have one pitch diameter. We want to ask two questions at this juncture. 1) Why do we care about pitch diameter? And 2) How do we measure it?

We care about pitch diameter for the same reason our gear associates do. It is in this area that the load our fasteners apply during thread engagement is found. Gear makers will spend time actually measuring the contact area at the pitch diameter to assure quality contact. This is their run to daylight. Good flank contact at the pitch line makes for long gear life. For those of us with screw threads in our blood, we want to know what the threads are doing at the pitch diameter. Let's say we have a pointer that we can use to touch on point right at the pitch diameter and then drag it across the cylinder to measure this distance. This would be the pitch diameter of the thread we just measured. We can see this in Figure 2.

Let's next take a set of five pointers, tied together, and attempt to place them at this same location. Unless our thread is perfect in geometry, the pointers will indicate a

Basic Thread Included Angle: A
Half Angle: a_1 and a_2

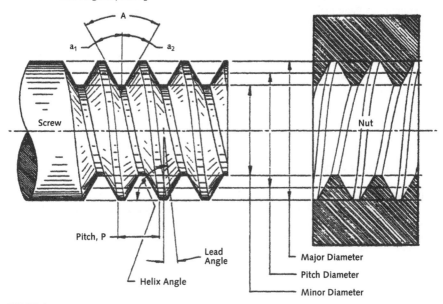

FIGURE 1.

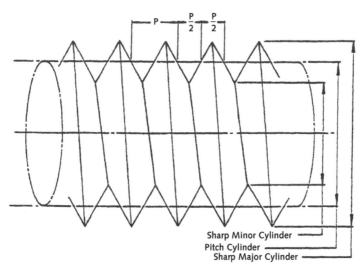

FIGURE 2. Pitch Cylinder
The pitch cylinder is one of such diameter and location of its axis that its surface would pass through a straight thread in such a manner as to make the widths of the thread ridge and the thread groove equal and, therefore, is located equidistantly between the sharp major and minor cylinders of a given thread form.

reading higher than the pitch diameter. The amount higher will be equal to the variation in geometry. A gage maker, working with toolmaker precision, will control slight variations in the geometry to four decimal places or better accuracy. Even these variations add up to make the diameter measured along five threads larger than the measure of just one. The five thread measure is actually the *Functional Size*. Functional size is pitch diameter plus thread variation. In equation form:

FUNCTIONAL SIZE =
PITCH DIAMETER +
CUMULATIVE VARIATION

This is useful for us to know because it is this differential that enables us to gage the goodness of contact between threads in application. ∎

SKILLKEEPER

A customer wants to use either 0.500-13 UN 2A or an M12 x 1.75 6g threaded stud to clamp a pneumatic cylinder head with a mating acorn nut. She would like to know how much each nut advances before seating for each quarter turn of the wrench. What are the distances?

SOLUTION

0.500-13:
0.25 (1/13 inches) = 0.0192 inches

M12 x 1.75:
0.25(1.75mm) = 0.44 mm

TECHNICALSUPPORT

• •

Running True

by Jim Speck, P.E.   Reprinted from *American Fastener Journal*, November/December 2005

I've found myself listening to a lot of doctors recently. Trying to understand something technical and removed from your knowledge base can be frustrating. What helps are knowledgeable people who speak honestly and plainly. Not an easy combination. The answers to your questions from these people run true.

The performance of machine screw threads can be technical and frustrating if you are trying to get maximum performance with minimum input. Thread dimensional information that consistently runs true can make a big difference. Let's consider a 3A bolt thread to be purchased and installed into a 2B-tapped hole. Like a patient making a medical decision, we have a choice of suppliers. How do we decide? Let's take a look at the final threaded assembly in Figure 1.

Clearly, even for a commodity-type machine screw thread, there are a dozen geometric factors for each part, and two dozen when you include the mating part, which determine whether our assembly enjoys *rude good health* or is *sick as a dog*. What average layperson is going to inspect every feature of every possible supplier? Not many. Fortunately, thread science gives us an answer to the "screw thread is better quality" question, which is honest and clear. First, let's intro-

duce a time-tested principle, known as the Taylor Principle:

> *"The maximum material limits of as many related elements or dimensions as practicable should be incorporated in the Go gage, whereas the minimum material limit of each related element or dimension may be gaged only by individual minimum material limit gages or gaging methods."*

Reread the Taylor Principle until you have it down solid. For both external and internal threads, we will use gage elements, which measure as many dimensions as practical. These elements are full form gages and their measurement gives us the Go or Maximum Material Size of each threaded part.

The Minimum Material Size is the individual thread pitch diameter measurement. Most often, this is best performed with Cone and Vee elements. See Figure 2 for Cone and

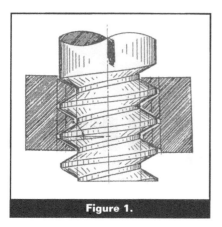

Figure 1.

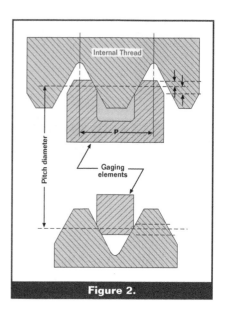

Figure 2.

Vee elements measuring the internal thread.

If any given screw thread is running really true, with no variation in any of the dimensions—in other words, if the thread is *perfect*, functional (Go) diameter measured data will exactly equal pitch diameter (no Go). Mathematically, we can say it like this:

EXTERNAL THREAD:

FUNCTIONAL DIAMETER =
PITCH DIAMETER + SUM of
Thread Variation

INTERNAL THREAD:

FUNCTIONAL DIAMETER =
PITCH DIAMETER – SUM of
Thread Variation

If thread variation is zero, functional diameter = pitch diameter. Since we won't require our threaded component suppliers to be perfect, merely *very good*, the answer to our thread quality question "Which parts have better thread quality?" is:

ANSWER: The ones with both measurements in specification and the smallest differential. Direct pitch diameter measurement is also well established as ANSI B1.3M, System 22. ∎

PROBLEM

A customer is using a .7500-20 UNEF 3A externally threaded shaft into a .7500-20 UNEF 2B Nut. What are the max. and min. limits for pitch and functional diameters?

SOLUTION

.7500-20 UNEF-3A	Max. .7175	Min. .7142
	Tolerance = .0033	
.7500-20 UNEF-2B	Max. .7218	Min. .7175
	Tolerance = .0043	

NOTE: The above is from a real application. Initial components had an interference issue during assembly. The problem was resolved on following shipments when both threads were measured using System 22 thread gaging. One part is manufactured in Asia, the other in New England.

Reprinted from *U.S. Fastener Report*, June 2012

ARE YOU CERTAIN?

by Jim Speck, P.E.

But how can you know for sure? This is a question we all think at some point in our lives. What and whom to trust. What is in our best interest. From childhood, and as a parent if we are so fortunate, having confidence that something "rings true" and is based upon a foundation of trust and sound experience. It starts with the old "roots and wings" thing—knowledge that we stand on a firm ground and confidence to expand our understanding into new areas.

With aerospace fasteners, the concept of *"certainty"* acquires a specific, technical definition. To be certain of a measurement, or test result of an aerospace fastener, or any mechanical component, implies a firm knowledge. A certainty of the number. Every such knowledge has its limit. We "know" the value of the force of gravity, until we are moving at gravity escaping velocity. We know the diameter of a screw thread at its pitch diameter and the depth of a drive recess, until we increase the level of our precision. As an example, a statistical process requirement once called for the recess depth to be recorded during a production run. Measurements were made with the agreed upon calibrated recess penetration gage. The operator, knowing the run was being monitored, changed header punches at the first sign of wear and the measurements were a consistent 0.026″. An X Bar R chart was filled with 15 groups of 5 measurements each, 75 measurements in all, of 0.026″. Not wanting to challenge the data, but doubting the numbers, I went out in the shop after closing and made a series of measurements. Sure enough, every one was 0.026″.

At a customer's SPC training class, I made a point about this chart, "with no variation." I learned an important lesson from the instructor, who said, "There is

variation, it is just at four decimal places and your inspection system's level of precision is only three."

Providing an aerospace inspection measurement without putting it into the framework of its certainty is leaving out an important part of the number. Let's take the pitch diameter of an SAE AS8879 UNJ external thread of a critical aerospace fastener. The uncertainty of our pitch diameter measurement is important for anyone using this measurement to make decisions with respect to the fastener. The uncertainty is a very useful tool to help anyone using this measurement to determine how much confidence to place in it when making a decision about the pitch diameter of this aerospace fastener, or the lot from which it is drawn.

As an example, let's select a .2500-20 UNJC-3A aerospace fastener screw to be inspected. We shall inspect two lots of a sample of 10 parts from each lot.

What is our certainty about these measurements? We could measure them again. Two points could be made:

1. If the gage system is accurate, repeated readings should be within the gage tolerance class precision. In our application that would be thread gage tolerance Class "X" which is .0003″ per Federal Standard H28, Section 6, Table 6.8.

Table 1 shows the data for the pitch diameters for the 20 parts, by lot.

TABLE 1. .2500-20 UNJC-3A Per SAE AS8879, Category I, Pitch Diameter Measurement Pitch Diameter: .2175" to .2147"		
	LOT 1	**LOT 2**
Part 1	.2156	.2168
2	.2154	.2172
3	.2150	.2173
4	.2151	.2170
5	.2154	.2167
6	.2156	.2170
7	.2150	.2167
8	.2148	.2164
9	.2150	.2172
10	.2151	.2170
Mean	.2152"	.2169"
Standard Deviation	.0003"	.0003"

2. For any specific reading, there is a range of possible readings. Let's pick Lot 1, Part #5, whose recorded pitch diameter reading is .2154" and examine this number.

Lot 1, Part #5's pitch diameter reading could be represented by the range shown in Table 2.

TABLE 2.	
Highest Probable Value	
Recorded Reading	.2154"
Lowest Probable Value	

This reading falls within the range of class X precision to four decimal places. We are confident of the .215" values. The .0004" could be within a range which, while very small, can vary depending on the uncertainty of our inspection system. Our system is comprised of the part, gage, inspector and the inspection environment.

There are two generally accepted methods for determining the numerical value of this uncertainty. The first uses an uncertainty budget which assigns values to all of the significant factors which influence the system. The second approach uses the informed judgment of the evaluator to assign an appropriate uncertainty factor.

It is this second method which can have much value for the fastener inspector. By having an understanding of the probable range around the "tenths" value of a measurement, judgment can be used for how close a reading is to a specification limit. Further use of the mean and standard deviation of recorded readings, often easy to obtain with scientific calculators and software programs, can increase the knowledge and judgment with respect to both specific types of aerospace fasteners, the measuring and testing equipment used and the inspection report data used by manufacturers, distributors and consumers of aerospace fasteners. ∎

Reprinted from *U.S. Fastener Report*, June 2012

INTELLECTUAL PROPERTY

Intellectual property. Just the phrase presents an image of important work. Its value carefully developed with great creative effort, time and money. And this can all be true. Intellectual property also applies to items we see everyday. Famous trademarks, designs of all kinds, from high-tech industrial to high fashion accessories.

THIS IS MINE, YOU CAN'T TAKE IT!

BY JIM SPECK, P.E.
JOHNSON GAGE COMPANY

The development, identification, promotion and protection of intellectual property is taking on renewed importance in today's global economy and our informational efficiency. That was brought home to me by an email from Mike McGuire a few weeks ago. At dinner in Scottsdale this winter, we were discussing the global fastener business and I mentioned the critical importance intellectual property will acquire as a competitive asset. A casual comment really, and one I didn't take much note of until Mike sent me a Federal Register notice and asked if I could elaborate. So here goes.

This will be a story of three friends: Mike McGuire, who started this; Dick Hrinak, with whom I've had the pleasure of enjoying more fastener business adventures than either of us can laugh about these many years later; and Janet Winkelman, who asked me a question following an MFDA after-dinner talk at the start of her fastener career, and a few years ago walked and talked me through a crushing weight of grief.

Mike started this discussion by framing it as "international fastener commerce," and then returned to it with impeccable timing. *That's Mike.*

In the fastener industry, there are all kinds of fastener designs that took creativity, hard work and someone's hard earned money to develop and grow into profitable fasteners. When fasteners make a profit, competitors want in and customers want to save some money. Nothing wrong with that. The question is, when does competition become stealing someone else's work? To answer Mike, I'd offer that work plus time equal the intellectual property value. Creative people tend to be more tuned in to them, but ideas are everywhere. That's the 1%. The 99% is refining, developing, testing and selling the idea. When this work is obvious to the market, it represents intellectual value which exists as an economic and legal fact of accountable presence.

But by whose standards does this accounting take place? Many decades ago, Japan was an economic juggernaut. Their industry was more efficient, they worked longer and harder, heck we were even told their kids were smarter than our kids. Why even bother to turn on the office lights in the morning? Working for Dick Hrinak, we realized that every culture has its competitive advantage. They were high quality and had a lower cost basis, but we could think up a fastener design and have it running on the headers and shipped to the customer before someone in Asia could read the RFQ. When we started seeing China's amazing economic growth, I asked Dick where he saw the advantage this time. Without missing a beat, he said "moral compass." In our culture, most of us are raised to believe a man's word is his bond. If we say we are going to do something, we do it. That is our culture, but not in every country. A very perceptive observation. *That's Dick.*

Intellectual property in international borders requires elevated care. As the owner of a design, increased effort in protecting and defending intellectual property is going to be needed. Let everyone, everywhere know where your boundaries are, and be ready, willing and able to defend them. Do not rely on every worldwide competitor playing by the same rules.

Intellectual property is, at its root, a very personal value—to the individual designer, their company and subsequent owners and marketers. Life is very often not fair. In discussing life's lessons with Janet, she listens as only a truly good person can. And one of the fundamental lessons we learn as we grow, is that people lie. We are taught as children to tell the truth, as the story about George Washington and the cherry tree. But in business and life, you learn it's not always going to be that way. I never saw that as clearly until I talked about life that day. It took the thoughtful conversation of a good friend to help me see that clearly. *That's Janet.*

In commercial exchanges with the market, international and domestic, realize the world does not owe you fairness. But it does work. When it comes to intellectual property, believe nothing and trust no one. Always get it in writing, in the languages of the cultures involved. Tell them what they need to know, but not everything you know. Defend your intellectual work with everything you've got. And always, always be working on the next profitable idea. •

TECHNICALSUPPORT

• •

by Jim Speck, P.E., The Johnson Gage Company

Reprinted from *American Fastener Journal*, January/February 2013

SYSTEMS

21 22 23

for Fastener Screw Threads

As I work around people involved in fastener screw thread inspection, a topic that comes up a lot is the place and merits of screw thread inspection systems 21, 22 and 23. Since I work for Johnson Gage, an assumption is that I'll have a bias toward the latter two systems. I attempt to provide a good technical answer. Usually the conversations become more relaxed and informative. I often learn as much as I pass on.

With this short piece, I will try to do the same. The systems used to be known as Method A, Method B and Method C, and it is interesting to look at how the numbers 21, 22 and 23 were assigned. I've yet to get an answer to that one. What is known is that System 21 is screw thread measurement by attributes. Systems 22 and 23 are thread measurement systems by variables. There is nothing inherently "good" or "bad" about either.

A good technical starting point is defining any given screw thread's maximum and minimum material conditions. Picture the side of a fastener's thread section (a cap screw's, for example). Then, make the screw thread shape a set of lines, like the dividing lines on a highway lane. The line on the outside is our maximum material condition, and the inside line is our minimum material condition. The screw thread's maximum material condition

is that size which, if we added one more tiny grain of material, would be outside the line and oversized.

The screw thread's minimum material condition would be that size which, if we removed one tiny grain of material, the screw thread would be undersized. When we inspect a screw thread, we are attempting to determine: Is this screw thread within or outside of the maximum and minimum material conditions for its specification?

A good way to visualize this is to see the part threads overlaid against these two lines in two dimensions, or as a cylindrical concentric shape in three dimensions. Systems 21, 22 and 23 all use the go gage to answer the maximum material side of the question. The go ring does this with a limit; the System 22 functional gage does this with a numerical value. If each is qualified, meaning the gage is set properly to a known standard reference,

System 22 External Thread Gage

this answer is true in either system.

The minimum material part of the answer to our question is the key difference between thread inspection systems. The System 21 no go, ring or plug establishes the no go functional limit. By design, it is not able to contact every surface of the thread. If the thread is within minimum material limits, this presents no problem. If, however, the thread has some area below the minimum material limit, our no go gage may not detect it. A good technical term for this is confounding.

The no go gage can be confounded. It has a higher level of uncertainty with respect to determining if a part's screw threads are within the minimum material limit.

A system 22 and 23 minimum material gage should provide the ability to directly contact any such area. The technical description we learn here at Johnson Gage is the Taylor Principle. It says that the gage used for the maximum material condition should contact all of the geometry that makes up a part's maximum material condition, and the gage used for the minimum material condition should isolate that feature. The feature we isolate with system 22 and 23 gages is the product's pitch diameter.

Variable screw thread gaging is sometimes called differential gaging, as the difference between the single element pitch and full form functional measurement values is an indicator of the thread form variance in the screw threads being inspected.

If the thread form is ideal (as near-perfect as possible), the differential approaches zero. Focusing the thread inspection on a fastener's pitch diameter seems to be very efficient. It looks to find those threads which might be a weak link in the part. During the banking crisis, there was a lot of talk about stress testing banks. Pitch diameter measurement is a dimensional stress test of the fastener's screw threads.

All three systems are valid and available. The decision of which to use is the business of the parties involved, with the party paying the invoice—the customer—rightly having the majority say.

My viewpoint is that it is good business to know and understand each system and make informed decisions in the direction that does the most good for any given application. ∎

Reprinted from *American Fastener Journal*, May/June 2014

TECHNICALSUPPORT

• • • • • • • • • • • • • •

ADVICE FOR A

Fastener Specialist

by Jim Speck, P.E.

A machinist's mirror is a very useful tool in mechanical work. It helps see the underside of cold header tools during set-up. You and I look in the mirror with much thought as we greet and work through each new day. When I look at the mirror now, I see an older man where once there was a younger one. If the old man could give the young one advice on fasteners, what good advice would it be?

I've given that some thought. Thinking back to all the good people who've taken time to share an insightful observation, technical point or a new way to look at a difficult problem, here is some of what I'd like to say:

1 Take a systems view. It's an irony that the screw manufacturer does no assembly in making fasteners, which only work as part of a system of screw, internally threaded part and fastened assembly. It is important to look at the entire system, not just the screw, in fastener application engineering. Define the application, set goals, and troubleshoot applications with a systems-wide viewpoint. One of my favorite examples is the Skyscraper Window application. A tall building had window glass vibrating loose and falling from their window frames soon after opening. The windows had been designed to resist wind loads higher than any recorded on that site. The problem presented a real public safety hazard. A model of the building had been wind tunnel tested, and the wind glass and frames should have held as designed. But when wind tunnel tests were performed with models of the

other buildings in the neighborhood, air turbulence developing additional wind peak forces was indicated. For the application specialist, know the entire system into which a fastener is to be installed.

2 Be a student for life. There was a time when books and sharing tribal knowledge were primary means of learning new skills and techniques. The information age has expanded one's options, but it also expanded the diversions. Be a careful consumer of information. For the fastener specialist, The *American Fastener Journal* is a good place to start. Read the articles and attend shows, conferences and presentations. Always be on the lookout to learn something useful and new. Consider it your personal continuous improvement program.

3 Be well rounded. A book that influenced me encouraged engineers to be "civil engineers." It told the story of famous, successful engineering, emphasizing the importance of being knowledgeable in more than just your specialty. I always admire the idea of the Renaissance man:

that individual who is learned in engineering, the arts, sport, business and civil duties. For the fastener specialist, follow your better interests in learning and developing new skills. As an undergraduate, I was required to take courses in public speaking and what was then called "engineering economics." Those two courses provided me with skills I have used profitably for a good life. You never know what you will learn early that you may really need to succeed somewhere down the road.

4 Share your learning. We are all, at different times, both the student and the teacher. Take time at both. When in the presence of anyone from whom you can learn, be a good student. Similarly, when presented, often by circumstance, with the opportunity to teach someone something new and useful, don't hesitate. There is an old saying: I taught them all they know, but not all I know. I like that. You don't need to divulge all of your tricks of the trade. That said, enthusiastically and freely spread the fundamentals. That is one of the ways that great industries like fastener businesses are built.

5 Make team members, not adversaries. When you are working an application problem, make the problem the focus. Work to get all involved pulling in the same direction—getting to the root cause and implementing an efficient and reliable solution. Don't make the same mistake twice. Learn from it, and respond proactively.

6 Put dollar values on projects. Don't do things just to do them. Fastener application is a business. Know what the economics of an application are, and work to have it make economic sense.

7 Keep things simple and efficient. It is easy to get overly technical just because it impresses people. Often, the better the solution, the more simple and obvious it is when put in place.

8 Numerical values rule. In fastener applications, putting numbers on measurements makes analysis better. Words are great for description, but numbers are best for fastener application work. Saying the threads are "tight" is not as clear as indicating a "two thousandths" or "two tenths" clearance between pressure flanks. The word "tight" could mean different things to different people. With "two thousandths" or "two tenths," we've become much more accurate.

9 Break applications into sections. For example, if you are evaluating a bolt, I would make the head, shank and screw threads separate sections for study. Practice the ability to understand each as a contributing component: The head as a torque input device and an applier of bearing force, the shank as an elongation and clamp force generating member, and the threads as converters of rotary into linear motion which is reversible and whose pressure flanks anchor the clamping load. Use the same application-analysis-by-sections technique for any fastener component you are evaluating.

10 Listen to understand and give back. As you encounter each application, understand what is required. READ the specifications on the requests for quote and purchase orders, and ask questions on any that are unclear. Engage in productive conversations, and share what you know.

In conclusion, work to be a productive member of your company, the fastener industry and society. We all gain when each of us develops more value than we consume.

With time comes experience. Know that not all are going to be good, but most offer ample opportunity to learn and grow. Sometimes the ones that turn out least well offer the most new knowledge. Fasteners hold our modern industrial, commercial and military world together. If you work at it diligently, you can spend a working lifetime studying and improving them and still learn something new every day. ■

Appendix E: Some Additional Problem Solution Ideas and Summary Thoughts

Following Appendix D, which was a large section of wide-ranging fastener articles written over many years, this concluding appendix is offered as a small section to add some thoughts, do some "housekeeping," and point the reader forward, if not to infinity, to at least "go long," scoring early and often in your fastener career.

Fastened assemblies can be incredibly complex. Their mathematical analysis may require a rigorous approach and computational requirements, and they are in all probability beyond the scope of this text and author. With this point firmly established, there is much value in an intelligent individual working in the field that is capable of performing first-pass calculations to provide guidance in supply, problem-solving, and continuously improving fastened products. When I took a preparatory course to prepare for the professional engineering examinations, one of the instructors, a licensed professional engineer and practicing professional, made the point that with all of the powerful and sophisticated stress analysis hardware and software available, an engineer should never lose the ability to do "back of the envelope calculations."

This ability makes decision-making a more intelligent and efficient process.

To reiterate from the previous, I have found three fundamental equations to prove a useful set of tools for this work:

1. $S = P/A$, the equation of stress, load, and area
2. $T = K*D*P$, the equation of tightening torque, fastener/component friction, nominal diameter, and load
3. $e = F*L/A*E$, the equation of elongation, load, grip length, stress area, and modulus of elasticity.

Here are some problems for the reader to think about with the goal of launching you on a path of your own continual fastener knowledge improvement program.

Draw a V and set with an included angle of 60°. This is a cross-section of a screw thread. The thread has a vertical height, which we call H, and a horizontal distance across the tops of the V, which we will call P. We have theoretical thread height and pitch. Screw thread dimensions are standardized on ratios of H and P. Now locate the point on the inner flanks where the distance across the inner part, the V groove, would be equal in horizontal distance to that across the inner flanks and to the distance across a ridge made with an attached V of the same P and H dimensions. This locates our pitch diameter datum. Finally, draw a circle of a diameter, which allows it to contact the flanks right at the pitch diameter datum. This is the concept providing a foundation for the best wire size of a screw thread of symmetrical flanks, which is used in the measurement over wires, sometimes called the three-wire method to measure screw thread gage threads in a metrology laboratory.

Imagine a flat round cylindrical plate with three arms starting at the center and carried outward at 0°, 120°, and 240° to the circumference of the plate. Find the increase in diameter if we add an equal distance X to each arm. Use this hint—this can be done through logical reasoning, graphically, or experimentally. This is based on an actual gaging problem.

Design a threaded fastener size with a safety factor that you select for a static load that you specify. Use the equation $S = P/A$, with A being the minimum thread stress area.

Calculate the increase in an external screw thread's functional diameter and the corresponding decrease in an internal screw thread's functional diameter for a plating or coating thickness of 0.006–0.008 mm. Use this hint—the ratio of plating thickness to functional diameter delta, idealized, is 1:4.

Plot the fit of pre- and postplate screw threads being assembled graphically. Use this hint—an I beam with a horizontal line at mid-I beam representing the maximum material condition works well.

Consider a cooling tower requirement for corrosion-resistant bolts, nuts, and washers. The fastener raw material is a given. Make a list of factors that could be used to increase the installed value of the fasteners you are supplying. Use this hint—separate the factors into categories of strength (static and dynamic) appearance, and reusability.

Creativity—Where do good fastener application ideas come from? Place yourself in a brainstorming mindset. On your first pass, no idea is "bad," get them down to paper. Then refine your list. Play mental catch with an associate tossing ideas back and forth. Be on the lookout, "observe," and value time on the shop floor. Read trade publications of the specific and related industries.

Make a simple free body diagram (FBD) of a single tensile force and reaction.

Read a bridge overpass and when convenient and safe, sketch it on paper. Make an FBD of what you estimate to be first the static loads and reactions including turning moments. Then superimpose dynamic forces, hurricane force wind and rain, and/or overweight large vehicles traveling rapidly.

Using a watch, clock, or timer as a "time gage," record the arrival times of, for example, people to a meeting or cars passing by. Consider the distribution of this inspection data. Can you make some inference from the population you inspected? Was it typical or were there significant differences? These are your process statistics. On your time gage and with you as its user, what factor could have influenced how good they were with respect to being accurate? These factors form the basis of your uncertainty "budget." Together with the date, it helps any user in using the data you have collected.

You are a member of a forensics team examining a gas valve that ruptured, breaking the bolts helping to fasten it. List what steps you would think to perform to help determine the cause and how to make the valve more robust. This is an example of what a chief engineer once assigned to several of us, including the author early in his fastening career.

In conclusion, I am reminded of the story of the old man and the star fish. The morning after a big storm at the shore of an ocean, an old man was walking on the beach, picking up star fish that had been tossed ashore, and were in risk of dying from drying out in the rising sun and unable to get back to the receding ocean tides. A young woman walking by observed the old man and said, "You are a fool, why waste your time and effort. You will never be able to make a difference to these star fish." The old man slowly bent, picked one up, and carefully tossed it back safely into the water. "It will make a difference to that one!" the old man replied.

Writing a book like this is not easy. There exists the risk that I will have made a poor use of time otherwise profitably engaged. It is my sincere hope that at least to one reader, it can be truthfully said it made a difference.

I wish you much productivity in your fastening, joining, and assembly applications. I also wish that you are fortunate to have the good company of many friends and find abundant friendship, enjoy your success, and travels.

Jim Speck
Avon, CT
June 14, 2014

Post Script: In many of the Fastener Fundamental Workshops I have conducted, there occurred at the conclusions a desire to go an extra mile in the closing minutes by providing some additional application information. In that spirit, I want to offer these:

Where do ideas come from? Cultivate the ability to think expansively about fastener solutions. Find locations and associates that help good fastener ideas. The practice of playing "mental catch" with a technical associate often helps.

When taking on a difficult task, make the project something you are fortunate to be able to work on and share your enthusiasm widely. When projects go well, celebrate and distribute the accolades to all productively involved in the work.

Be the champion of your projects to help them overcome obstacles.

When using math, remember equal has to be equal. Have integrity, be able to show your work and ask to see that of others.

Go the extra kilometer. Work a little harder and for a while longer than you want at jobs worth doing.

Take time to laugh, especially with loved ones and build lasting memories.

Index

A

A 286 stainless alloy, 210–211, 318; *see also* Stainless steel
AC, *see* Alternating current (AC)
Acceleration, 75; *see also* Fastener dynamics
Accessibility, 123–124
Adhesive, 60–61; *see also* Nonmetal joints assembly, 94–95
Aerospace, 167
After-service work, 15
AISI, *see* American Iron and Steel Institute (AISI)
Allen heads, 170
Alternating current (AC), 226
Alternator, 227
Aluminum, 211, 215; *see also* Fastener materials; Nonferrous materials alloys, 213
American Iron and Steel Institute (AISI), 53; *see also* Steel
AISI 1010 steel, 200
AISI 1018 steel, 201–202
AISI 1038 carbon steel, 202
AISI 4037 carbon alloy steel, 203
AISI 4140 steel, 203–204
AISI 52100 ball-bearing steel, 205
AISI 8640 medium carbon screw, 204–205
American National Standards Institute (ANSI), 18, 126
American Society for Testing and Materials (ASTM), 126
American Society of Mechanical Engineers (ASME), 18
Amplitude, 154
Anaerobic adhesive, 287; *see also* Fasteners
Angle of inclination, 32
Angstroms, 103
Angular motion, 75; *see also* Fastener dynamics

ANSI, *see* American National Standards Institute (ANSI)
Anti-seizing, 171
Anvil, 123
Arc, 211, 213
Area, 277
thread, 9
ASME, *see* American Society of Mechanical Engineers (ASME)
Assembly, 115, 281; *see also* Build quantities; Circular fastener areas; Economic factors in fastener/assembly decisions; Engineering economics; Fasteners; Fastener workshops; Legal considerations and constraints
analysis, 249
anvil, 123
bedding-in effect, 21
efficiency estimation, 266–267
mistake, 294–295
model, 116, 117
planned obsolescence, 122
robustness, 122–124
tooling, 186
tree diagram, 298
Assembly sites and systems, 167; *see also* Automatic assembly machines; Fastener numerical methods; Modern assembly; Reusability factors; Simple input/output diagram
accessibility, 167–168
assembly training, 179–180
degree of automation, 180–184
design vs. servicing balance, 169
mating fastening clamping surfaces, 168
smart machines and robots, 187–189
torque transducer-driven wrenching system, 188